INSTRUCTOR'S MANUAL

CHEMISTRY

PRINCIPLES AND REACTIONS

THIRD EDITION

MASTERTON ▪ HURLEY

WILLIAM L. MASTERTON

UNIVERSITY OF CONNECTICUT

SAUNDERS GOLDEN SUNBURST SERIES

Saunders College Publishing
Harcourt Brace College Publishers

Fort Worth Philadelphia San Diego New York Orlando Austin
San Antonio Toronto Montreal London Sydney Tokyo

PREFACE

This manual starts off with a section entitled "Lecture Outlines",
which you may find helpful in adapting the text to your class schedule.
Beyond that, the manual is organized by text chapters. For each chapter,
we include three different features:

1. "Lecture Notes", which indicate the amount of time that we devote
to each chapter and the topics that we emphasize. Included are detailed
lecture outlines, which may serve as a guide for your own lectures. At a
minimum, they indicate how we cover topics and how successive topics can
be integrated.

2. A list of demonstrations illustrating topics covered in the chapter.
These are taken from three sources:

- the manual "Tested Demonstrations in Chemistry" (1994), Volumes I
and II, compiled and edited by George Gilbert by arrangement with the
Journal of Chemical Education. These are coded as "GILB" with the experiment
number (e.g., M 12).

- "Chemical Demonstrations" (1983-1992), Volumes 1-4, published by
Bassam Shakhashiri with many collaborators and contributors. These are
listed as "SHAK" followed by the volume and page reference.

- demonstrations described in the Journal of Chemical Education, mostly
since 1989, with the journal reference.

In addition, video demonstrations from three different sources are
listed:

- Shakhashiri and Schreiner (SHAK), 1991, Saunders College Publishing.

- Journal of Chemical Education Software (JCES) catalog 1996 (demonstration videos listed on p. 23).

- Falcon Software catalog 1996 (demonstration videos listed on p. 11).

3. Answers and detailed solutions to:
    - text problems which do not have answers in Appendix 6
    - challenge problems at the end of each chapter

Note that the problems in the text that are numbered in color are
answered in Appendix 6. Detailed solutions to these problems are given
in the "Solutions Manual" written by Prof. Cassandra Eagle and available
from Saunders College Publishing.

TABLE OF CONTENTS

LECTURE OUTLINES

This text, unlike other general chemistry texts now available, can be covered in its entirety in a one-year course. A reasonable schedule is shown below; further comments on the amount of time you should devote to each chapter are given in the body of this manual. It is assumed for convenience that you are teaching 14-week semesters with two 50-minute lectures per week. If three class periods are devoted to examinations each semester, that leaves 25 for covering material. On that basis, you should be able to complete Chapter 10 (Solutions) in the first semester. The second semester will then start with Reaction Rates (Chapter 11).

FIRST SEMESTER SCHEDULE

Week	Lecture	Topic
1	1	Chapter 1 (Matter and Measurements)
	2	Chapter 1
2	3	Chapter 2 (Atoms, Molecules, and Ions)
	4	Chapter 2
3	5	Chapter 3 (Mass Relations in Chemistry)
	6	Chapter 3
4	7	Chapter 3
	8	EXAM I
5	9	Chapter 4 (Reactions in Aqueous Solution)
	10	Chapter 4
6	11	Chapter 4
	12	Chapter 5 (Gases)
7	13	Chapter 5
	14	Chapter 6 (Electronic Structure)
8	15	Chapter 6
	16	Chapter 6
9	17	EXAM II
	18	Chapter 7 (Covalent Bonding)
10	19	Chapter 7
	20	Chapter 7
11	21	Chapter 8 (Thermochemistry)
	22	Chapter 8
12	23	Chapter 9 (Liquids and Solids)
	24	Chapter 9
13	25	Chapter 9
	26	EXAM III
14	27	Chapter 10 (Solutions)
	28	Chapter 10

Saunders College Publishing

SECOND SEMESTER SCHEDULE

Week	Lecture	Topic
1	1	Chapter 11 (Rate of Reaction)
	2	Chapter 11
2	3	Chapter 11
	4	Chapter 12 (Gaseous Chemical Equilibrium)
3	5	Chapter 12
	6	Chapter 13 (Acids and Bases)
4	7	Chapter 13
	8	Chapter 13
5	9	EXAM I
	10	Chapter 14 (Acid-Base Equilibria in Solution)
6	11	Chapter 14
	12	Chapter 15 (Complex Ions)
7	13	Chapter 15
	14	Chapter 16 (Precipitation Equilibria)
8	15	Chapter 17 (Spontaneity of Reaction)
	16	Chapter 17
9	17	EXAM II
	18	Chapter 18 (Electrochemistry)
10	19	Chapter 18
	20	Chapter 18
11	21	Chapter 19 (Nuclear Reactions)
	22	Chapter 20 (Chemistry of the Metals)
12	23	Chapter 20
	24	EXAM III
13	25	Chapter 21 (Chemistry of the Nonmetals)
	26	Chapter 21
14	27	Chapter 22 (Organic Chemistry)
	28	Chapter 22

If you want to use lecture time for review, for going over assigned problems, or for doing a large number of demonstrations, you will have trouble keeping up with this schedule. As you've almost certainly learned by now, the solution to this problem is not to talk faster; judicious deletions work better. It's been said, and wisely, that the secret of giving a good lecture is knowing what to leave out. Possible candidates include:

- introductory material on matter in Chapter 1 and atomic theory in Chapter 2. The chances are your students have been exposed to this material more than once in high school and understood it reasonably well the first time.

- Boyle's and Charles' laws in Chapter 5. We start the chapter by writing the ideal gas law and go on from there.

- the First Law discussion in Chapter 8. Quite frankly, this has very little to do with chemistry; students will not be irreparably damaged if they are unaware of the distinction between H and E.

- the discussion of colligative properties in Chapter 10 could be shortened; Raoult's law could easily be omitted.

Saunders College Publishing

    - reaction mechanisms in Chapter 11. Students have a lot of trouble with this; we're not sure it's worth the effort.
    - polyprotic acids in Chapter 14.
    - qualitative analysis in Chapter 16, particularly if it does not fit with your laboratory schedule.

Beyond these selective omissions, some instructors may want to delete one or another of the descriptive chapters at the end of the text (Ch. 20-22). If, in that way, you can squeeze out a couple of lectures, they can well be spent on Chapter 12 (three lectures instead of two) and Chapter 19 (two lectures instead of one).

Textbook authors sometimes tell you that chapters can be covered in almost any order, depending on your preference. That isn't really true for this textbook or any other with structural integrity. It can be done, but only with very careful additions and deletions of material. Suppose, for example, you want to cover Precipitation Equilibria (Chapter 16) immediately after Acid-Base Equilibrium (Chapter 14). Keep in mind that an understanding of formation constants (Complex Ions, Chapter 15) is assumed when methods of dissolving precipitates are considered in Section 16.2 of Chapter 16.

Saunders College Publishing

CHAPTER 1
Matter and Measurements

LECTURE NOTES

This material ordinarily requires 2 lectures (100 min), allowing for a 10-15 minute introduction to the course in the first lecture. If you're in a hurry, this can be cut to $1\frac{1}{2}$ lectures by discussing only quantitative material (significant figures, unit conversions, density, solubility).

A few points to keep in mind:

. - virtually all of your students will be familiar with the metric system and prefixes. It may be worth discussing the basis of SI, but you don't have to dwell on it.

- students readily learn the rules of significant figures but typically ignore them after Chapter 1. It may help to emphasize that these are common-sense (albeit approximate) rules for estimating experimental error.

- many students (typically the weaker ones) stubbornly resist using conversion factors, preferring instead a rote method with which they became infected in high school. It may be useful to point out that conversion factors will be used throughout the course and hence should be learned at this point.

- students often have trouble with solubility calculations. The approach used in the text involves conversions (Example 1.8). The solubility is considered to be a conversion factor relating grams of solute to grams of solvent.

<u>LECTURE 1</u>

<u>I Types of Substances</u>

   A. <u>Elements</u> Cannot be broken down into simpler substances. Examples: nitrogen, lead, sodium, arsenic. Symbols: N, Pb, Na, As
   B. <u>Compounds</u> Contain two or more elements with fixed mass percents. Sodium chloride: 39.34% Na, 60.66% Cl
     Glucose: 40.00% C, 6.71% H, 53.29% O

C. <u>Mixtures</u> Homogeneous (solutions) vs heterogeneous. Separation by filtration, distillation.

II <u>Measured Quantities</u>

   A. <u>Length</u> Base unit is the meter. $1 \text{ km} = 10^3 \text{ m}$; $1 \text{ cm} = 10^{-2} \text{ m}$; $1 \text{ mm} = 10^{-3} \text{ m}$; $1 \text{ nm} = 10^{-9} \text{ m}$. Dimensions of very tiny particles will be expressed in nanometers.

   B. <u>Mass</u> $1 \text{ kg} = 10^3 \text{ g}$; $1 \text{ mg} = 10^{-3} \text{ g}$. Two different kinds of balances will be used in lab. Analytical balance (± 0.001 g) should be used only for accurate, quantitative work.

   C. <u>Temperature</u> $t_{\circ F} = 1.8 t_{\circ C} + 32°$; $T_K = t_{\circ C} + 273.15$

      Convert 68°F to °C and K:

$$t_{\circ C} = (68° - 32°)/1.8 = 20°C; \quad T_K = 293$$

   D. <u>Derived Units</u>
      1. <u>Volume</u> $1 \text{ L} = 10^3 \text{ mL} = 10^3 \text{ cm}^3 = 10^{-3} \text{ m}^3$
      2. Energy Joule = energy consumed when 10-watt bulb burns for 0.1 second. $4.184 \text{ J} = 1 \text{ cal}$. Burning match evolves about 0.5 kcal = 2 kJ

LECTURE 2

I <u>Experimental Error; Significant Figures</u>

Suppose object is weighed on crude balance to ±0.1 g and mass is found to be 23.6 g. This quantity contains 3 "significant figures", i.e., 3 experimentally meaningful digits. With an analytical balance, mass might be 23.582 g (5 sig. fig.).

   A. <u>Counting Significant Figures</u>
      1. Volume of liquid = 24.0 mL; three significant figures. Zeros at end of measured quantity, following nonzero digits, are significant.
      2. Volume = 0.0240 L; three significant figures (note that 0.0240 L = 24.0 mL). Zeros at beginning of a measured quantity, preceding nonzero digits, are not significant.
   B. <u>Multiplication and Division</u> Keep only as many significant figures as there are in the least precise quantity. Density of piece of metal weighing 36.123 g with volume of 13.4 mL?

$$\text{density} = \frac{36.123 \text{ g}}{13.4 \text{ mL}} = 2.70 \text{ g/mL}$$

   C. <u>Addition and Subtraction</u> Keep only as many digits after the decimal point as there are in the least precise quantity. Add 1.223 g of sugar to 154.5 g of coffee:

$$\text{Total mass} = 1.2 \text{ g} + 154.5 \text{ g} = 155.7 \text{ g}$$

Note that rule for addition and subtraction does not relate to significant figures. Number of significant figures often decreases upon subtraction:

$$\begin{array}{lll}
\text{Mass beaker + sample} = 52.169\ \text{g} & & 5\ \text{sig. fig.} \\
\text{Mass empty beaker} = 52.120\ \text{g} & & 5\ \text{sig. fig.} \\
\text{Mass sample} = 0.049\ \text{g} & & 2\ \text{sig. fig.}
\end{array}$$

   D. <u>Exact Numbers</u> "one liter" means 1.00000 . . . L

II <u>Conversion Factors</u>

   A. <u>Simple, one-step conversions</u>
     1. A rainbow trout is measured to be 16.2 in long. Length in centimeters?

$$\text{length in cm} = 16.2\ \text{in} \times \frac{2.54\ \text{cm}}{1\ \text{in}} = 41.1\ \text{cm}$$

Note cancellation of units. To convert from centimeters to inches, would use the conversion factor 1 in/2.54 cm.

2. Barometric pressure reported on Canadian radio to be 99.6 kPa. Express in mm Hg (101.3 kPa = 760 mm Hg)

$$\text{pressure (mm Hg)} = 99.6\ \text{kPa} \times \frac{760\ \text{mm Hg}}{101.3\ \text{kPa}} = 747\ \text{mm Hg}$$

Method is particularly useful with unfamiliar units.

   B. <u>Multiple Conversion Factors</u> Baseball thrown at rate of 89.6 miles per hour. Speed in meters per second?

$$1\ \text{mile} = 1.609\ \text{km} = 1.609 \times 10^3\ \text{m};\ 1\ \text{h} = 3600\ \text{s}$$

$$\text{speed} = 89.6\ \frac{\text{mile}}{\text{hr}} \times \frac{1.609 \times 10^3\ \text{m}}{1\ \text{mile}} \times \frac{1\ \text{hr}}{3600\ \text{s}} = 40.0\ \text{m/s}$$

III <u>Properties of Substances</u>

Distinguish between intensive vs extensive, chemical vs physical

   A. <u>Density</u> Empty flask weighs 22.138 g. Pipet 5.00 mL of octane into flask; total mass = 25.598 g

$$d = 3.460\ \text{g}/5.00\ \text{mL} = 0.692\ \text{g/mL}$$

Volume occupied by ten grams of octane?

$$V = 10.00\ \text{g} \times \frac{1\ \text{mL}}{0.692\ \text{g}} = 14.5\ \text{g}$$

B. <u>Solubility</u> Often expressed as grams of solute per 100 g of solvent. Example:

temperature	10°C	100°C
soly. lead nitrate (g/100g water)	50	140

1. How much water is required to dissolve 80 g of lead nitrate at 100°C?

$$\text{Mass water} = 80 \text{ g lead nitrate} \times \frac{100 \text{ g water}}{140 \text{ g lead nitrate}} = 57 \text{ g water}$$

2. Cool to 10°C. How much lead nitrate remains in solution?

$$\text{Mass lead nitrate} = 57 \text{ g water} \times \frac{50 \text{ g lead nitrate}}{100 \text{ g water}}$$

$$= 28 \text{ g lead nitrate}$$

$$80 \text{ g} - 28 \text{ g} = 52 \text{ g lead nitrate crystallizes}$$

DEMONSTRATIONS

1. Scientific method: GILB H 29
2. Reaction of iron with sulfur: GILB A 6, SHAK <u>1</u> 55
3. Decomposition of mercury(II) oxide: GILB A 8
4. Separation of a mixture: GILB A 14
5. Reaction of sodium with chlorine: GILB A 24, A 25, SHAK <u>1</u> 61
6. Chromatography: GILB Q 3, Q 13
7. Density of liquids: GILB C 13, SHAK <u>3</u> 229
8. Significant figures: J. Chem. Educ. <u>69</u> 497 (1992)

VIDEO

1. Reaction of sodium with chlorine: Falcon 16
2. Miscibility and density of liquids: SHAK 36

PROBLEMS

1. a. mixture   b. compound   c. mixture   d. element

3. a. solution   b. heterogeneous   c. heterogeneous

5. a. mass; grams or ounces   b. volume; liters or quarts

c. pressure; pounds per square inch

7. a. $303 \text{ m} \times \dfrac{1 \text{ km}}{10^3 \text{ m}} = 0.303 \text{ km} < 303 \times 10^3 \text{ km}$

b. $500 \text{ g} \times \dfrac{1 \text{ kg}}{10^3 \text{ g}} = 0.500 \text{ kg} = 0.500 \text{ kg}$

c. $1.50 \text{ cm}^3 \times \dfrac{1 \text{ m}^3}{10^6 \text{ cm}^3} \times \dfrac{10^{27} \text{ nm}^3}{1 \text{ m}^3} = 1.50 \times 10^{21} \text{ nm}^3 > 1.50 \times 10^3 \text{ nm}^3$

9. $t_{°C} = \dfrac{68° - 32°}{1.8} = 20°C; \quad T = 293 \text{ K}$

11. $t_{°C} = 308° - 273° = 35°C; \quad t_{°F} = 1.8(35°) + 32° = 95°F$

13. a. $1 \text{ L} = 10^{-3} \text{ m}^3$ b. $1 \text{ J} = 1 \text{ kg·m}^2/\text{s}^2$

c. $1 \text{ kPa} = 10^3 \text{ Pa} = 10^3 \text{ kg/m·s}^2$

15. a. $3.65 \times 10^5 \text{ kPa} \times \dfrac{10^3 \text{ Pa}}{1 \text{ kPa}} \times \dfrac{1 \text{ atm}}{1.013 \times 10^5 \text{ Pa}} = 3.60 \times 10^3 \text{ atm}$

b. $375 \text{ J} \times \dfrac{1 \text{ kJ}}{10^3 \text{ J}} \times \dfrac{1 \text{ kcal}}{4.184 \text{ kJ}} = 0.0896 \text{ kcal}$

17. a. 3   b. 4   c. 4   d. 4   e. 1

19. $V = \pi (2.500 \text{ cm})^2 \times 1.20 \text{ cm} = 23.6 \text{ cm}^3$

21. a. 0.528 g/cm^3 b. 2.71 mi/hr c. 1481 g

d. $461 \text{ g}/5.6 \text{ cm}^3 = 82 \text{ g/cm}^3$

23. a. 7.49 g b. 298.69 cm c. 13 lb d. $3.5 \times 10^2 \text{ oz}$

25. a. $22.3 \text{ mL} \times \dfrac{1 \text{ L}}{10^3 \text{ mL}} = 2.23 \times 10^{-2} \text{ L}$

b. $22.3 \text{ cm}^3 \times \dfrac{1 \text{ in}^3}{(2.54 \text{ cm})^3} = 1.36 \text{ in}^3$

c. $2.23 \times 10^{-2} \text{ L} \times \dfrac{1.057 \text{ qt}}{1 \text{ L}} = 2.36 \times 10^{-2} \text{ qt}$

27. a. 27 fardells $\times \dfrac{1 \text{ nooke}}{2 \text{ fardell}} \times \dfrac{1 \text{ yd}}{4 \text{ nooke}} \times \dfrac{1 \text{ kide}}{4 \text{ yd}} = 0.84$ kide

   b. 15 fardells $= \dfrac{15}{32}$ kide;

     $\left(\dfrac{15}{32} \text{ kide}\right) \times (12 \text{ kide}) \times \dfrac{16 \text{ yd}^2}{1 \text{ kide}^2} = 9.0 \times 10^1 \text{ yd}^2$

29. 1 acre $\times \dfrac{1 \text{ hectare}}{2.47 \text{ acre}} \times \dfrac{10^4 \text{ m}^2}{1 \text{ hectare}} \times \dfrac{(39.37 \text{ in})^2}{1 \text{ m}^2} \times \dfrac{1 \text{ ft}^2}{144 \text{ in}^2} = 4.36 \times 10^4 \text{ ft}^2$

31. 1 pt $\times \dfrac{1 \text{ qt}}{2 \text{ pt}} \times \dfrac{1 \text{ L}}{1.057 \text{ qt}} = 0.473$ L ; $\dfrac{0.473}{6.0} \times 100\% = 7.9\%$

33. $\dfrac{235 \text{ kJ}}{250 \text{ mL}} \times \dfrac{1 \text{ kcal}}{4.184 \text{ kJ}} \times \dfrac{10^3 \text{ mL}}{1 \text{ L}} \times \dfrac{1 \text{ L}}{1.057 \text{ qt}} \times \dfrac{1 \text{ qt}}{4 \text{ cups}} = 53.1$ kcal/cup

35. 27.0 g $\times 0.900 \times \dfrac{16 \text{ oz}}{453.6 \text{ g}} \times \dfrac{\$5.18}{1 \text{ oz}} = \$4.44$

Worth more as source of silver

37. 112 g/75.0 mL $=$ 1.49 g/mL

39. 12.65 g/6.9 mL $=$ 1.8 g/mL

41. $V = 11 \text{ in} \times 12 \text{ in} \times \dfrac{(2.54 \text{ cm})^2}{1 \text{ in}^2} \times t = 8.9 \text{ g}/(2.70 \text{ cm}^3)$

   Solving: $t = 3.9 \times 10^{-3} \text{ cm} \times \dfrac{10 \text{ mm}}{1 \text{ cm}} = 0.039$ mm

43. 10.0 L $\times \dfrac{10^3 \text{ mL}}{1 \text{ L}} \times \dfrac{1.01 \text{ g}}{1 \text{ mL}} \times 0.0500 = 5.05 \times 10^2$ g

45. a. 46.5 g water $\times \dfrac{37.0 \text{ g KCl}}{100 \text{ g water}} = 17.2$ g KCl

   b. 27.9 g KCl $\times \dfrac{100 \text{ g water}}{48.3 \text{ g KCl}} = 57.8$ g water

   c. 75.0 g water $\times \dfrac{37.0 \text{ g KCl}}{100 \text{ g water}} = 27.8$ g KCl ; no

     75.0 g water $\times \dfrac{48.3 \text{ g KCl}}{100 \text{ g water}} = 36.2$ g KCl; yes

47. a. physical   b. physical   c. physical   d. chemical

49. a. chemical property relates to behavior in reaction

    b. distillation involves vaporizing liquid; filtration takes off solid

    c. solute is a component of a solution

51. Pb: $1 \text{ g} \times \dfrac{1 \text{ cm}^3}{11.34 \text{ g}} = 8.818 \times 10^{-2} \text{ cm}^3$

    O: $1 \text{ g} \times \dfrac{1 \text{ cm}^3}{1.81 \times 10^{-3} \text{ g}} = 552 \text{ cm}^3$

    Lead is much denser than oxygen

53. $108 \text{ carat} \times \dfrac{2.00 \times 10^{-1} \text{ g}}{1 \text{ carat}} \times \dfrac{1 \text{ lb}}{453.6 \text{ g}} = 4.76 \times 10^{-2} \text{ lb}$

    $21.6 \text{ g} \times \dfrac{1 \text{ cm}^3}{3.51 \text{ g}} \times \dfrac{1 \text{ in}^3}{(2.54 \text{ cm})^3} = 0.376 \text{ in}^3$

55. $V = \pi r^2 (6.18 \text{ cm}) = 147.3 \text{ g} \times \dfrac{1 \text{ cm}^3}{4.55 \text{ g}}$

    Solving: $r^2 = 1.667 \text{ cm}^2$; diameter $= 2.58 \text{ cm}$

57. $2t_{\circ C} = 1.8 t_{\circ C} + 32°$

    $0.2 t_{\circ C} = 32°$; $t_{\circ C} = 160°$; $t_{\circ F} = 320°$

58. $31.5 \text{ gal} \times \dfrac{4 \text{ qt}}{1 \text{ gal}} \times \dfrac{1 \text{ L}}{1.057 \text{qt}} \times \dfrac{10^{-3} \text{ m}^3}{1 \text{ L}} \times \dfrac{1 \text{ km}^3}{10^9 \text{ m}^3} = (1.0 \times 10^{-10} \text{ km})(\text{area})$

    Solving: area $= 1.2 \text{ km}^2$

59. $V = \dfrac{10.0 \text{ g}}{2.70 \text{ g/cm}^3} = \pi (0.254 \text{ cm})^2 \times \ell$; $\ell = 18.3 \text{ cm}$

60. $\dfrac{8.50 \times 10^3 \text{ L}}{\text{d}} \times \dfrac{1 \text{ m}^3}{10^3 \text{ L}} \times \dfrac{7.0 \times 10^{-6} \text{ g Pb}}{1 \text{ m}^3} \times 0.75 \times 0.50 \times \dfrac{365.2 \text{ d}}{1 \text{ yr}}$

    $= 8.1 \times 10^{-3} \text{ g Pb}$

CHAPTER 2
Atoms, Molecules, and Ions

LECTURE NOTES

Students find this material relatively easy to assimilate; it's almost entirely qualitative. On the other hand, there's a lot of memorizing (sorry: learning) involved. This chapter is readily covered in two lectures, perhaps $1\frac{1}{2}$. Some general observations:

- material in Sections 2.1-2.3 is generally well-covered in high school chemistry courses; you need not dwell on it.

- students need to learn the molecular formulas of the elements (Figure 2.8), charges of transition metal cations (Figure 2.10), names and formulas of polyatomic ions (Table 2.2) and the nomenclature system for oxoacids.

- naming compounds requires students to distinguish ionic vs molecular substances. It helps to point out that binary molecular compounds are composed of two nonmetals. Almost all ionic compounds contain a metal cation.

- the periodic table will be discussed in greater detail later in the text. The same is true of nuclear reactions (Section 2.4).

<u>LECTURE 1</u>

I <u>Atoms</u>

  A. <u>Atomic theory</u>
    Postulates: Elements consist of tiny particles called atoms, which retain their identity in reactions. In a compound, atoms of two or more elements are combined in a fixed ratio of small whole numbers (e.g., 1:1, 2:1, 3:2, etc.).

  B. <u>Components</u>

	relative mass	relative charge	location
proton	1	+1	nucleus
neutron	1	0	nucleus
electron	0.0005	-1	outside

  C. <u>Atomic number</u> = number of protons in nucleus = number of electrons in neutral atom. Characteristic of a particular element; all H atoms have 1 proton, all He atoms have 2 protons, etc.

D. <u>Mass number</u> = number of protons + no. of neutrons. Atoms of same element can differ in mass number.

	prot.	neut.	at. no.	nucl. symbol	mass no.
carbon-12	6	6	6	$^{12}_{6}C$	12
carbon-14	6	8	6	$^{14}_{6}C$	14

$^{12}_{6}C$ and $^{14}_{6}C$ are referred to as isotopes

II <u>Periodic Table</u>

A. <u>Structure</u> Periods and groups; numbering system for groups

B. <u>Metals</u> located at lower left of table, nonmetals at upper right. Metalloids

III <u>Nuclear Reactions</u>

Certain unstable nuclei decay spontaneously by

A. α-<u>emission</u> (loss of He-4 nucleus)

$$^{238}_{92}U \longrightarrow {}^{4}_{2}He + {}^{234}_{90}Th$$

B. β-<u>emission</u> (loss of electron; neutron $\longrightarrow$ proton + electron)

$$^{234}_{90}Th \longrightarrow {}^{234}_{91}Pa + {}^{0}_{-1}e$$

C. γ-<u>emission</u>; not included in nuclear equation

LECTURE 2

I <u>Molecules</u>

A. Usually made up of nonmetal atoms; held together by covalent bonds.

B. <u>Types of formulas</u> Consider the compound ethyl alcohol:

Molecular formula: C_2H_6O

<pre>
                    H   H
                    | |
Structural formula: H - C - C - O - H
                    | |
                    H   H
</pre>

Condensed structural formula: CH_3CH_2OH

II <u>Ions</u>

A. <u>Formation of monatomic ions</u>

Na atom (11 p^+, 11 e^-) $\rightarrow$ Na$^+$ ion (11 p^+, 10 e^-) + e^-

F atom (9 p^+, 9 e^-) + e^- $\rightarrow$ F$^-$ ion (9 p^+, 10 e^-)

Nucleus remains unchanged

B. <u>Charges of monatomic ions</u>
1. Ions with noble gas structures
Cations: Group 1 (+1); Group 2 (+2); Al^{3+}
Anions: Group 16 (−2); Group 17 (−1); N^{3-}
2. Transition metal cations (Figure 2.10)

C. <u>Polyatomic ions</u>
Names and formulas (Table 2.2)

D. <u>Formulas of Compounds</u> Apply principle of electrical neutrality

calcium fluoride: Ca^{2+}, F^- ions: CaF_2

aluminum nitrate: Al^{3+}, NO_3^-: $Al(NO_3)_3$

sodium dihydrogen phosphate: Na^+, $H_2PO_4^-$: NaH_2PO_4

III <u>Names of Compounds</u>

A. <u>Ionic</u> Name cation, followed by anion. Note that with transition metal cations, charge is indicated by Roman numeral.

Na_2SO_4 sodium sulfate $Fe(NO_3)_3$ iron(III) nitrate

Note that, except for ammonium salts, ionic compounds contain metal cations.

Systematic names of oxoanions (ate, ite; per, hypo)

calcium hypochlorite: $Ca(ClO)_2$

B. <u>Binary molecular compounds</u> Use of Greek prefixes

SF_6 sulfur hexafluoride N_2O_3 dinitrogen trioxide

C. <u>Acids</u>
binary acids: hydrochloric acid
oxo acids: ate salt $\rightarrow$ ic acid; ite salt $\rightarrow$ ous acid
$HClO_4$ perchloric acid $HClO$ hypochlorous acid

DEMONSTRATIONS

1. Law of constant composition: GILB A 12

2. Law of conservation of mass: GILB A 16

3. Simulation of Rutherford's experiment: GILB L 7

4. Isotope effects (H_2O and D_2O): GILB M 18

5. Reaction of hydrogen with chlorine: GILB H 38

6. Reaction of iron with chlorine: SHAK $\underline{1}$ 66

VIDEO

1. Cathode rays: Falcon 1

2. Electron in a magnetic field: Falcon 2

3. Isotopes, heavy water ice cubes: JCES 3

4. Reaction of hydrogen with chlorine: JCES 19

5. Sodium chloride crystal cleavage: JCES 4

PROBLEMS

1. No change in mass in chemical reaction

3. a. none   b. Conservation of Mass   c. Constant Composition

    d. Multiple Proportions

5. 8 O–□ , 3 □–□

7. % O = $\dfrac{3.87}{52.30}$ x 100% = 7.40%;   % O = $\dfrac{1.25}{16.93}$ x 100% = 7.38%

    same within experimental error

9. a. A: 0.875   B: 0.438   C: 1.75

    b. 1.75/0.875 = 2;   1.75/0.438 = 4

11. a. Rutherford   b. see p. 29

13. $^{80}_{34}$Se

15. a. yes   b. no; mass number of second isotope unknown

17. a. isobars: Ca-41, K-41, Ar-41; isotopes: Ca-40, Ca-41

    b. same atomic number (20)   c. same number of P + n

19. a. 26   b. 30   c. 26   d. 30, 26, 23

14

21. 0, 7, 7, 7; $^{37}_{17}Cl^-$, -1; $^{56}_{27}Co^{2+}$, 25; -2, 34, 48, 36

23. a. 60 x 6 = 360 p$^+$, 360 e$^-$ b. 6 p$^+$, 10 e$^-$

c. 10 p$^+$, 10 e$^-$ d. 1 p$^+$, 0 e$^-$

25. a. Cs   b. W   c. Sb   d. P   e. K

27. a. metal   b. metal   c. metalloid   d. nonmetal   e. metal

29. a. 4 (with #111)   b. 6   c. 6 (including H)   d. 4

31. $^{210}_{82}Pb \rightarrow {}^{210}_{83}Bi + {}^{0}_{-1}e$; beta

33. a. $^{230}_{90}Th \rightarrow {}^{4}_{2}He + {}^{226}_{88}Ra$ b. $^{210}_{82}Pb \rightarrow {}^{210}_{83}Bi + {}^{0}_{-1}e$

c. $^{187}_{75}Re \rightarrow {}^{187}_{76}Os + {}^{0}_{-1}e$ d. $^{247}_{100}Fm \rightarrow {}^{4}_{2}He + {}^{243}_{98}Cf$

35. a. $^{235}_{92}U$ b. $^{231}_{91}Pa$

37. $^{218}_{84}Po$, $^{214}_{82}Pb$, $^{210}_{80}Hg$; $^{210}_{81}Tl$, $^{210}_{82}Pb$

39. a. $^{12}_{6}C$ b. $^{6}_{3}Li$ c. $^{14}_{7}N$

41. all except Th-225

43. a. CH_3COOH, $C_2H_4O_2$ b. C_2H_5Cl, C_2H_5Cl

45. a. NH_3 b. CO_2 c. C_2H_2 d. XeF_4 e. $GeCl_4$

47. a. xenon trioxide   b. hydrazine   c. tetrasulfur teranitride
d. carbon tetrachloride   e. bromine pentafluoride

49. NaI, Na_2O, BaI_2, BaO

51. a. $K_2Cr_2O_7$ b. $NaNO_3$ c. $Mg_3(PO_4)_2$ d. CrO   e. $NiCl_3$

53. a. iron(III) acetate   b. sodium chlorate   c. potassium hydrogen
                                                              carbonate
d. cadmium(II) sulfide   e. potassium oxide

55. $BrCl_3$, perchloric acid, nitrogen trifluoride, $KMnO_4$,
nickel(II) hydrogen phosphate, SO_2

57.

Saunders College Publishing

59. a. usually   b. usually   c. usually

61. a. $SiCl_4$; silicon tetrachloride

 b. $Ca_3(PO_4)_2$; calcium phosphate

 c. $Al_2(SO_3)_3$; aluminum sulfite

63. a. ethane: 18.0 g C/4.53 g H = 3.97 g C/g H
  ethene: 43.20 g C/7.25 g H = 5.96 g C/g H
  5.96/3.97 = 1.50 = 3/2

 b. CH_2 and CH_3; C_2H_4 and C_2H_6; - -

64. mass = $13(1.6727 \times 10^{-24} g) + 13(9.1095 \times 10^{-28} g)$
  $+ 14(1.6750 \times 10^{-24} g) = 4.5207 \times 10^{-23}$ g

 $V = \dfrac{4\pi}{3}(1.43 \times 10^{-8} cm)^3 = 1.22 \times 10^{-23} cm^3$

 $d = 4.5207 \ g/1.22 \ cm^3 = 3.71 \ g/cm^3$

 Empty space between Al atoms

65. 1.4965×10^{-23} g $- 2(9.1095 \times 10^{-28}$ g$) = 1.4963 \times 10^{-23}$ g

66. a. $200 \times 500 \times 2.5 \times 10^{19}$ molecules = 2.5×10^{24}

 b. $(2.5 \times 10^{24})/(1.1 \times 10^{44}) = 2.3 \times 10^{-20}$

 c. $(2.3 \times 10^{-20}) \times 500 \times (2.5 \times 10^{19}) \approx 2.9 \times 10^2$

CHAPTER 3
Mass Relations in Chemistry; Stoichiometry

LECTURE NOTES

This chapter is considerably more difficult and time-consuming than the two that precede it. It contains a considerable amount of quantitative material that is fundamental for future chapters. we suggest you devote three lectures to Chapter 3. The first lecture deals with atomic masses and the mole (Sections 3.1, 3.2), the second with the quantitative aspects of chemical formulas (Section 3.3), the third with mass relations in reactions (Section 3.4). Points to keep in mind include the following:

1. Students have little trouble calculating atomic masses from isotopic data. The reverse process, estimating isotopic abundances from atomic masses, is much more difficult because it involves solving an algebraic equation. Incidentally, the treatment of significant figures here is a bit tricky because the percent abundances are dependent on one another. In general, the atomic mass can be calculated with more precision than you might suppose at first glance from abundance data.

2. Note that mole-gram conversions (Section 3.2) will be required in many later chapters, often as the first step in a more complex problem.

3. When dealing with formulas (Section 3.3), it is important to emphasize early on that the subscripts give not only the atom ratio but also the mole ratio. It is necessary that students realize this if they are to follow the logic of obtaining simplest formulas from mass percents.

4. Students ordinarily have little trouble calculating formulas from mass percents. They are much less adept at obtaining formulas from analytical data such as that given in Example 3.9.

5. It is important to get across the point (Section 3.4) that a chemical equation describes what happens when a reaction is carried out in the laboratory. Including in the equation the physical states of reactants and products helps to emphasize this point.

6. When you discuss mass relations in reactions, some students will

revert to the infamous "ratio-and-proportion" method. The comments made
in Chapter 1 about conversion factors apply here too.

7. There are many different ways to find the limiting reactant and
calculate the theoretical yield. We've tried most of them and recommend
the approach described in Section 3.4.

LECTURE 1

I <u>Atomic and Formula Masses</u>

    A. <u>Meaning of atomic masses</u> - give relative masses of atoms. Based
on C-12 scale; most common isotope of carbon is assigned an atomic
mass of exactly 12 amu.

element	B	Ca	Ni
atomic mass	10.81 amu	40.08 amu	58.69 amu

A nickel atom is $58.69/40.08 = 1.464$ times as heavy as a calcium
atom. It is $58.69/10.81 = 5.429$ times as heavy as a boron atom.

    B. <u>Atomic masses from isotopic composition</u>

A. M. = (A. M. isotope 1)(%/100) + (A.M. isotope 2)(%/100) + - -

Isotope	Atomic mass	Percent
Ne-20	20.00 amu	90.92
Ne-21	21.00 amu	0.26
Ne-22	22.00 amu	8.82

A.M. Ne = $20.00(0.9092) + 21.00(0.0026) + 22.00(0.0882) = 20.18$ amu

    C. <u>Masses of individual atoms</u> Since the atomic masses of H, Cl and
Ni are 1.008 amu, 35.45 amu and 58.69 amu, it follows that

        1.008 g H, 35.45 g Cl, 58.69 g Ni

all contain the same number of atoms, N_A. It turns out that:

$$N_A = \text{Avogadro's number} = 6.022 \times 10^{23}$$

1. Mass of H atom:

$$1 \text{ atom H} \times \frac{1.008 \text{ g H}}{6.022 \times 10^{23} \text{ atoms}} = 1.674 \times 10^{-24} \text{ g}$$

2. Number of atoms in one gram of nickel:

$$1.000 \text{ g Ni} \times \frac{6.022 \times 10^{23} \text{ atoms Ni}}{58.69 \text{ g Ni}} = 1.026 \times 10^{22} \text{ atoms}$$

    D. <u>Formula mass</u> = sum of atomic masses in formula

FM H_2O = 18.02 amu   FM NaCl = 58.44 amu

If formula represents molecule, formula mass = molecular mass

II <u>The Mole</u>

    A. <u>Meaning</u> $1 \text{ mol} = 6.022 \times 10^{23}$ items

        $1 \text{ mol H} = 6.022 \times 10^{23}$ H atoms; mass $= 1.008$ g

        $1 \text{ mol Cl} = 6.022 \times 10^{23}$ Cl atoms; mass $= 35.45$ g

        $1 \text{ mol Cl}_2 = 6.022 \times 10^{23}$ Cl_2 molecules; mass $= 70.90$ g

        $1 \text{ mol HCl} = 6.022 \times 10^{23}$ HCl molecules; mass $= 36.46$ g

    B. <u>Molar mass</u>   Generalizing from the above examples, the molar mass, $\mathcal{M}$, is numerically equal to the formula mass.

	formula mass	molar mass
$CaCl_2$	110.98 amu	110.98 g/mol
$C_6H_{12}O_6$	180.18 amu	180.18 g/mol

LECTURE 2

I <u>The Mole</u>

    A. <u>Mole-mass conversions</u>
       1. Calculate mass in grams of 13.2 mol of $CaCl_2$

$$\text{mass} = 13.2 \text{ mol CaCl}_2 \times \frac{110.98 \text{ g CaCl}_2}{1 \text{ mol CaCl}_2} = 1.47 \times 10^3 \text{ g}$$

       2. Calculate number of moles in 16.4 g of $C_6H_{12}O_6$

$$\text{no. moles} = 16.4 \text{ g } C_6H_{12}O_6 \times \frac{1 \text{ mol } C_6H_{12}O_6}{180.18 \text{ g } C_6H_{12}O_6} = 0.0910 \text{ mol}$$

II <u>Formulas</u>

    A. <u>Mass percent from formula</u>   Percent composition of K_2CrO_4?
      molar mass $= (78.20 + 52.00 + 64.00)$g/mol $= 194.20$ g/mol

      $\% \text{ K} = \dfrac{78.20}{194.20} \times 100\% = 40.27\%$ $\% \text{ Cr} = \dfrac{52.00}{194.20} \times 100\% = 26.78\%$

      $\% \text{ O} = \dfrac{64.00}{194.20} \times 100\% = 32.96\%$

      Note that percents must add to 100

    B. <u>Simplest formula from percent composition</u>
      Find mass of each element in sample of compound.   Then find numbers of moles of each element and finally the mole ratio.

      Simplest formula of compound containing 26.6% K, 35.4% Cr, 38.0% O?

      Work with 100 g sample.   26.6 g K, 35.4 g Cr, 38.0 g O

no. moles K = 26.6 g $\times$ $\dfrac{1 \text{ mol}}{39.10 \text{ g}}$ = 0.680 mol K

no. moles Cr = 35.4 g $\times$ $\dfrac{1 \text{ mol}}{52.00 \text{ g}}$ = 0.681 mol Cr

no. moles O = 38.0 g $\times$ $\dfrac{1 \text{ mol}}{16.00 \text{ g}}$ = 2.38 mol O

Note that   2.38/0.680 = 3.50 = 7/2   Simplest formula: $K_2Cr_2O_7$

C. <u>Simplest formula from analytical data</u>
A sample of acetic acid (C, H, O atoms) weighing 1.000 g burns to give 1.466 g CO_2 and 0.6001 g H_2O. Simplest formula?
Find mass of C in sample (from CO_2), then mass of H (from H_2O) and finally mass of O by difference.

mass C = 1.466 g CO_2 $\times$ $\dfrac{12.01 \text{ g C}}{44.01 \text{ g } CO_2}$ = 0.4001 g C

mass H = 0.6001 g H_2O $\times$ $\dfrac{2.02 \text{ g H}}{18.02 \text{ g } H_2O}$ = 0.0673 g H

mass O = 1.000 g - 0.400 g - 0.067 g = 0.533 g

no. moles C = 0.4001 g C $\times$ $\dfrac{1 \text{ mol C}}{12.01 \text{ g C}}$ = 0.0333 mol C

no. moles H = 0.0673 g H $\times$ $\dfrac{1 \text{ mol H}}{1.008 \text{ g H}}$ = 0.0666 mol H

no. moles O = 0.533 g O $\times$ $\dfrac{1 \text{ mol O}}{16.00 \text{ g O}}$ = 0.0333 mol O

simplest formula is CH_2O

D. <u>Molecular formula from simplest formula</u>
Must know molar mass. For acetic acid, molar mass = 60 g/mol
Formula mass = 30 amu; 60/30 = 2
Molecular formula: $C_2H_4O_2$

<u>LECTURE 3</u>

I <u>Chemical Equations</u>

A. <u>Balancing</u> Must have same number of atoms of each type on both sides. Achieve this by adjusting coefficients in front of formulas. Example: combustion of propane in air to give carbon dioxide and water.

$$C_3H_8(g) + O_2(g) \longrightarrow CO_2(g) + H_2O(l)$$

Balance C: $\qquad C_3H_8(g) + O_2(g) \longrightarrow 3CO_2(g) + H_2O(l)$

Balance H: $\qquad C_3H_8(g) + O_2(g) \longrightarrow 3CO_2(g) + 4H_2O(l)$

Balance O: $\qquad C_3H_8(g) + 5O_2(g) \longrightarrow 3CO_2(g) + 4H_2O(l)$

Meaning: 1 mol C_3H_8 reacts with 5 mol O_2 to form 3 mol CO_2 and 4 mol H_2O

B. <u>Mass relations in reactions</u>

   1. Moles of CO_2 produced when 1.65 mol C_3H_8 burns? Use coefficients of balanced equation to obtain conversion factor:

$$1.65 \text{ mol } C_3H_8 \times \frac{3 \text{ mol } CO_2}{1 \text{ mol } C_3H_8} = 4.95 \text{ mol } CO_2$$

   2. Mass of O_2 required to react with 12.0 g of C_3H_8?

$$12.0 \text{ g } C_3H_8 \times \frac{1 \text{ mol } C_3H_8}{44.09 \text{ g } C_3H_8} \times \frac{5 \text{ mol } O_2}{1 \text{ mol } C_3H_8} \times \frac{32.00 \text{ g } O_2}{1 \text{ mol } O_2} = 43.6 \text{ g } O_2$$

II <u>Yield of Product in Reaction</u>

A. <u>Limiting reactant, theoretical yield.</u> Ordinarily, reactants are not present in the exact ratio required for reaction. Instead, one reactant is in excess; some of it is left when the reaction is over. The other, limiting reactant, is completely consumed to give the theoretical yield of product

    To calculate the theoretical yield and identify the limiting reactant:

1. Calculate the yield expected if the first reactant is limiting.

2. Repeat this calculation for the second reactant.

3. The theoretical yield is the smaller of these two quantities. The reactant that gives the smaller calculated yield is the limiting reactant.

$$2Ag(s) + I_2(s) \longrightarrow 2AgI(s)$$

Calculate the theoretical yield of AgI and determine the limiting reactant starting with 1.00 g of Ag and 1.00 g of I_2.

theor. yield if Ag is limiting:

$$1.00 \text{ g Ag} \times \frac{469.54 \text{ g AgI}}{215.74 \text{ g Ag}} = 2.18 \text{ g AgI}$$

theor. yield if I_2 is limiting:

21

$$1.00 \text{ g } I_2 \times \frac{469.54 \text{ g AgI}}{253.80 \text{ g } I_2} = 1.85 \text{ g AgI}$$

Theoretical yield = 1.85 g AgI; I_2 is limiting

B. <u>Actual yield, percent yield</u>

$$\% \text{ yield} = \frac{\text{Actual Yield}}{\text{Theoretical Yield}} \times 100$$

Suppose actual yield of AgI were 1.50 g:

$$\% \text{ yield} = \frac{1.50}{1.85} \times 100 = 81.1$$

DEMONSTRATIONS

1. Relative masses of atoms (analogy): GILB L 2
2. Combustion of propane: GILB H 4
3. Theoretical yield: GILB A 11
4. Limiting reactant: GILB A 17, A 18
5. Reaction of antimony with halogens: GILB M 238; SHAK <u>1</u> 64
6. Hydrates of cobalt(II) chloride: J. Chem. Educ. <u>68</u> 779 (1991)

VIDEO

1. Reaction of metals with iodine: JCES 32
2. Reaction of aluminum with iodine: Falcon 18

PROBLEMS

1. b < a < c < d

3. 0.9976(16.00 amu) + 0.0004(17.00 amu) + 0.0020(18.00 amu)
    = 16.00 amu

5. 79.90 amu = 78.9440 amu(0.5057) + X(0.4943; X = 80.88 amu

7. 28.0855 = 27.977x + 28.977(0.9704 − x) + 29.974(0.0296)

    x = 0.920; 92.0% Si-28, 5.0% Si-29

9. Tall peak at mass number 24; much smaller and approximately equal peaks at 25 and 26.

11. a. 6.022×10^{23}

   b. $6.022 \times 10^{17}/20 = 3.011 \times 10^{16}$

13. a. $\dfrac{107.9 \text{ g}}{6.022 \times 10^{23}} = 1.792 \times 10^{-22} \text{ g}$

   b. $1.000 \times 10^{-12} \text{ g} \times \dfrac{6.022 \times 10^{23}}{107.9 \text{ g}} = 5.581 \times 10^{9}$ atoms

15. a. $0.285 \text{ mol} \times 207.2 \text{ g/mol} = 59.1 \text{ g}$

   b. $0.285 \text{ g} \times \dfrac{6.022 \times 10^{23} \text{ atoms}}{207.2 \text{ g}} = 8.28 \times 10^{20}$ atoms

   c. $\dfrac{0.285 \text{ g}}{207.2 \text{ g/mol}} \times 82 = 0.113 \text{ mol e}^{-}$

17. a. 54 b. $6.022 \times 10^{23} \times 54 = 3.252 \times 10^{25}$

   c. $5.0 \times 10^{3} \text{ g} \times \dfrac{1 \text{ mol}}{131.3 \text{ g}} \times 6.022 \times 10^{23} \times 54 = 1.2 \times 10^{27}$

19. a. 183.8 g/mol b. 48.00 g/mol

   c. $\mathcal{M} = 20(12.01 \text{ g/mol}) + 30(1.008 \text{ g/mol}) + 16.00 \text{ g/mol}$
   $= 286.44 \text{ g/mol}$

21. a. $\mathcal{M} = 9(12.01 \text{ g/mol}) + 8(1.008 \text{ g/mol}) + 4(16.00 \text{ g/mol})$
   $= 180.15 \text{ g/mol}$

   $0.1000 \text{ g} \times \dfrac{1 \text{ mol}}{180.15 \text{ g}} = 5.5509 \times 10^{-4} \text{ mol}$

   b. $\mathcal{M} = 27(12.01 \text{ g/mol}) + 46(1.008 \text{ g/mol}) + 16.00 \text{ g/mol}$
   $= 386.64 \text{ g/mol}$

   $1.000 \times 10^{-3} \text{ g} \times \dfrac{1 \text{ mol}}{386.64 \text{ g}} = 2.5864 \times 10^{-6} \text{ mol}$

   c. $\mathcal{M} = 12.01 \text{ g/mol} + 4(35.45 \text{ g/mol}) = 153.81 \text{ g/mol}$

   $2.95 \text{ g} \times \dfrac{1 \text{ mol}}{153.81 \text{ g}} = 0.0192 \text{ mol}$

23. a. $12.00 \text{ mol} \times 30.97 \text{ g/mol} = 371.6 \text{ g}$

b. $4(371.6 \text{ g}) = 1487 \text{ g}$

c. $\mathcal{M} = 30.97 \text{ g/mol} + 3.02 \text{ g/mol} = 33.99 \text{ g/mol}$

$33.99 \text{ g/mol} \times 12.00 \text{ mol} = 407.9 \text{ g}$

25. $\mathcal{M} = 3(12.01 \text{ g/mol}) + 6(1.008 \text{ g/mol}) + 16.00 \text{ g/mol} = 58.08 \text{ g/mol}$

a. $6.026 \times 10^{-3} \text{ mol}$, 3.629×10^{21}, 1.089×10^{22}

b. 155.9 g, 1.617×10^{24}, 4.851×10^{24}

c. $7.6 \times 10^{-10} \text{ g}$, $1.3 \times 10^{-11} \text{ mol}$, 2.4×10^{13}

d. $4.50 \times 10^{-8} \text{ g}$, $7.75 \times 10^{-10} \text{ mol}$, 4.67×10^{14}

27. $\mathcal{M} = 10(12.01 \text{ g/mol}) + 19(1.008 \text{ g/mol}) + 6(16.00 \text{ g/mol})$
$+ 30.97 \text{ g/mol} + 64.14 \text{ g/mol} = 330.36 \text{ g/mol}$

$\% \text{ C} = \dfrac{120.1}{330.36} \times 100 = 36.35; \quad \% \text{ H} = \dfrac{19.15}{330.36} \times 100 = 5.797$

$\% \text{ O} = \dfrac{96.00}{330.36} \times 100 = 29.06; \quad \% \text{ P} = \dfrac{30.97}{330.36} \times 100 = 9.375$

$\% \text{ S} = \dfrac{64.14}{330.36} \times 100 = 19.42$

29. $\mathcal{M} = 195.1 \text{ g/mol} + 28.02 \text{ g/mol} + 6.05 \text{ g/mol} + 70.90 \text{ g/mol}$
$= 300.1 \text{ g/mol}$

$250.0 \text{ mg} \times \dfrac{195.1}{300.1} = 162.5 \text{ mg Pt}$

31. a. $\dfrac{250.0}{611} \times 100 = 40.9\%$

b. $\mathcal{M} = 96.08 \text{ g/mol} + 9.07 \text{ g/mol} + 14.01 \text{ g/mol}$
$+ 32.00 \text{ g/mol} = 151.16 \text{ g/mol}$

$0.2500 \text{ g} \times \dfrac{14.01}{151.16} = 0.02317 \text{ g N}$

33. $\% \text{ C} = \dfrac{1.407 \text{ g CO}_2 \times 12.01 \text{ g C}/44.01 \text{ g CO}_2 \times 100}{1.000 \text{ g}} = 38.40$

$\% \text{ H} = \dfrac{0.134 \text{ g H}_2\text{O} \times 2.016 \text{ g H}/18.02 \text{ g H}_2\text{O} \times 100}{1.000 \text{ g}} = 1.50$

% Cl = $\dfrac{0.5228}{1.000}$ x 100 = 52.28

% O = 100.00 - 38.40 - 1.50 - 52.28 = 7.82

35. n Sb = $\dfrac{2.643 \text{ g}}{121.8 \text{ g/mol}}$ = 0.0217; n O = $\dfrac{0.521 \text{ g}}{16.00 \text{ g/mol}}$ = 0.0326

0.0217/0.0326 = 0.667; Sb_2O_3

37. Work with 100-g samples

a. n Al = $\dfrac{30.92 \text{ g}}{26.98 \text{ g/mol}}$ = 1.146 mol; n O = $\dfrac{45.87 \text{ g}}{16.00 \text{ g/mol}}$ = 2.867 mol

n H = $\dfrac{2.889 \text{ g}}{1.008 \text{ g/mol}}$ = 2.866 mol; n Cl = $\dfrac{20.32 \text{ g}}{35.45 \text{ g/mol}}$ = 0.5732 mol

1.146/0.5732 = 2; 2.867/0.5732 = 5 $Al_2O_5H_5Cl$

b. n C = $\dfrac{35.51 \text{ g}}{12.01 \text{ g/mol}}$ = 2.957 mol; n H = $\dfrac{4.77 \text{ g}}{1.008 \text{ g/mol}}$ = 4.73 mol

n O = $\dfrac{37.85 \text{ g}}{16.00 \text{ g/mol}}$ = 2.366 mol; n N = $\dfrac{8.29 \text{ g}}{14.01 \text{ g/mol}}$ = 0.592 mol

n Na = $\dfrac{13.60 \text{ g}}{22.99 \text{ g/mol}}$ = 0.5916 mol

2.957/0.5916 = 5; 4.73/0.5916 = 8; 2.366/0.5916 = 4

$$C_5H_8O_4NNa$$

c. n H = $\dfrac{2.24 \text{ g}}{1.008 \text{ g/mol}}$ = 2.22 mol; n C = $\dfrac{26.68 \text{ g}}{12.01 \text{ g/mol}}$ = 2.221 mol

n O = $\dfrac{71.08 \text{ g}}{16.00 \text{ g/mol}}$ = 4.442 mol CHO_2

39. mass C = 12.24 g x $\dfrac{12.01}{44.01}$ = 3.340 g; n C = 0.2781 mol

mass H = 2.505 g x $\dfrac{2.016}{18.02}$ = 0.2802 g; n H = 0.2780 mol

mass O = 5.287 g - 3.340 g - 0.280 g = 1.667 g; n O = 0.1042 mol

0.2781/0.1042 = 2.67 ; $C_8H_8O_3$

41. % C = $\dfrac{28.73 \text{ x } 12.01/44.01}{10.94}$ x 100 = 71.67

25

Saunders College Publishing

$$\% \text{ H} = \frac{5.386 \times 2.016/18.02}{10.94} \times 100 = 5.51$$

$$\% \text{ O} = 100.00 - 71.67 - 5.51 - 6.96 = 15.86$$

In 100 g of compound:

n C = 71.67/12.01 = 5.968 mol; n H = 5.51/1.008 = 5.47 mol

n N = 6.963/14.01 = 0.4970 mol; n O = 15.86/16.00 = 0.991 mol

5.968/0.497 = 12; 5.47/0.497 = 11; 0.991/0.497 = 2 $C_{12}H_{11}O_2N$

43. $$\% \text{ C} = \frac{14.36 \times 12.01/44.01}{6.315} \times 100 = 62.05$$

$$\% \text{ H} = \frac{7.832 \times 2.016/18.02}{6.315} = 13.88$$

$$\% \text{ N} = 100.00 - 62.05 - 13.88 = 24.07$$

mass C = 116.2 g x 0.6205 = 72.10 g = 6 mol

mass H = 116.2 g x 0.1388 = 16.12 g = 16 mol

mass N = 116.2 g x 0.2407 = 27.97 g = 2 mol

$$C_3H_8N; \ C_6H_{16}N_2$$

45. $\mathcal{M}$ = 24.30 g/mol + 32.07 g/mol + 64.00 g/mol + 126.11 g/mol

    = 246.48 g/mol

$$7.834 \text{ g} \times \frac{120.37}{246.48} = 3.826 \text{ g} \ ; \quad \% = \frac{126.11}{246.48} \times 100 = 51.164$$

47. $M_2 + MX_2 \longrightarrow M_3X_2$

49. a. $2Al(s) + Fe_2O_3(s) \longrightarrow Al_2O_3(s) + 2Fe(s)$

    b. $6CO_2(g) + 6H_2O(l) \longrightarrow C_6H_{12}O_6(s) + 6 \ O_2(g)$

    c. $C_2H_8N_2(s) + 2N_2O_4(g) \longrightarrow 3N_2(g) + 2CO_2(g) + 4H_2O(g)$

51. a. $Ba(s) + S(s) \longrightarrow BaS(s)$

    b. $Ba(s) + Br_2(l) \longrightarrow BaBr_2(s)$

    c. $Ba(s) + I_2(s) \longrightarrow BaI_2(s)$

    d. $3Ba(s) + N_2(g) \longrightarrow Ba_3N_2(s)$

    e. $2Ba(s) + O_2(g) \longrightarrow 2BaO(s)$

Saunders College Publishing

53. a. $2F_2(g) + H_2O(1) \rightarrow OF_2(g) + 2HF(g)$

b. $4NH_3(g) + 7\ O_2(g) \rightarrow 4NO_2(g) + 6H_2O(1)$

c. $2NH_4NO_3(s) \rightarrow 2N_2(g) + 4H_2(g) + 3\ O_2(g)$

d. $3Mg(s) + N_2(g) \rightarrow Mg_3N_2(s)$

e. $Au_2S_3(s) + 3H_2(g) \rightarrow 2Au(s) + 3H_2S(g)$

55. a. 32.6 mol O_2 b. 33.9 mol H_2O c. 13.9 mol PH_3

d. 0.220 mol P_4O_{10}

57. a. 7.88 mol PH_3 x $\dfrac{1\text{ mol }P_4O_{10}}{4\text{ mol }PH_3}$ x $\dfrac{283.9\text{ g }P_4O_{10}}{1\text{ mol }P_4O_{10}}$ = 559 g P_4O_{10}

b. 7.65 mol H_2O x $\dfrac{8\text{ mol }O_2}{6\text{ mol }H_2O}$ x $\dfrac{32.00\text{ g }O_2}{1\text{ mol }O_2}$ = 326 g O_2

c. 4.69 g H_2O x $\dfrac{1\text{ mol }H_2O}{18.02\text{ g }H_2O}$ x $\dfrac{4\text{ mol }PH_3}{6\text{ mol }H_2O}$ x $\dfrac{33.99\text{ g }PH_3}{1\text{ mol }PH_3}$

= 5.90 g PH_3

d. 12.42 g PH_3 x $\dfrac{1\text{ mol }PH_3}{33.99\text{ g }PH_3}$ x $\dfrac{2\text{ mol }O_2}{1\text{ mol }PH_3}$ x $\dfrac{32.00\text{ g }O_2}{1\text{ mol }O_2}$

= 23.39 g O_2

59. a. $SiO_2(s) + 2C(s) \rightarrow Si(s) + 2CO(g)$

b. 25.00 g Si x $\dfrac{1\text{ mol Si}}{28.09\text{ g Si}}$ x $\dfrac{1\text{ mol }SiO_2}{1\text{ mol Si}}$ = 0.8900 mol SiO_2

c. 32.55 g Si x $\dfrac{1\text{ mol Si}}{28.09\text{ g Si}}$ x $\dfrac{2\text{ mol CO}}{1\text{ mol Si}}$ x $\dfrac{28.01\text{ g CO}}{1\text{ mol CO}}$ = 64.91 g CO

61. a. $(7.50 \times 10^3 \times 0.045)$ g C_2H_5OH x $\dfrac{1\text{ mol }C_2H_5OH}{46.07\text{ g }C_2H_5OH}$ x $\dfrac{1\text{ mol }C_6H_{12}O_6}{2\text{ mol }C_2H_5OH}$

x $\dfrac{180.2\text{ g }C_6H_{12}O_6}{1\text{ mol }C_6H_{12}O_6}$ = 6.60×10^2 g glucose

b. $\dfrac{7.50 \times 10^3 \times 0.045}{46.07}$ mol CO_2 x $\dfrac{44.01\text{ g }CO_2}{1\text{ mol }CO_2}$ x $\dfrac{1\text{ L}}{1.80\text{ g }CO_2}$

= 1.8×10^2 L

Saunders College Publishing

63. $(1.00 \times 10^6 \times 0.012)$ g S $\times \dfrac{64.07 \text{ g SO}_2}{32.07 \text{ g S}} \times \dfrac{1 \text{ L}}{2.60 \text{ g SO}_2} = 9.2 \times 10^3$ L

65. $N_2(g) + 3H_2(g) \longrightarrow 2NH_3(g)$

   afterwards: 2 NH_3, 2 H_2, 4 N_2

67. a. $2Fe(s) + 3Cl_2(g) \longrightarrow 2FeCl_3(s)$

   b. Fe: 2.75 mol $FeCl_3$ Cl_2: 2.33 mol $FeCl_3$; Cl_2 is limiting

   c. 2.33 mol   d. 2.75 mol − 2.33 mol = 0.42 mol

69. a. $3Si(s) + 2N_2(g) \longrightarrow Si_3N_4(s)$

   b. 185 g $Si_3N_4(s) \times \dfrac{84.27 \text{ g Si}}{140.31 \text{ g Si}_3N_4} \times \dfrac{1}{0.87} = 1.3 \times 10^2$ g Si

71. a. C_2H_2: 125 g $\times \dfrac{176.04 \text{ g CO}_2}{52.07 \text{ g}} = 423$ g CO_2

   O_2: 125 g $\times \dfrac{176.04 \text{ g CO}_2}{160.00 \text{ g}} = 138$ g CO_2 = theor. yield

   b. $\dfrac{(60.1 \times 1.80) \text{g}}{138 \text{ g}} \times 100 = 78.4\%$

   c. 125 g $O_2 \times \dfrac{52.07 \text{ g C}_2H_2}{160.00 \text{ g O}_2} = 40.7$ g C_2H_2;   84 g unused

73.   mass PI_3 = 165 g $H_3PO_3 \times \dfrac{411.7 \text{ g PI}_3}{81.99 \text{ g H}_3PO_3} \times \dfrac{1}{0.550} = 1.51 \times 10^3$ g PI_3

   mass H_2O = 165 g $H_3PO_3 \times \dfrac{54.05 \text{ g H}_2O}{81.99 \text{ g H}_3PO_3} \times \dfrac{1.40}{0.550} = 277$ g H_2O (277 mL H_2O)

75. b. Mg   c. Mg   d. acid   e. (4)   f. 122 mL; 10 mL

77. $\mathcal{M}$ = (22.99 + 84.07 + 5.04 + 32.00)g/mol = 144.1 g/mol

   3.00 oz $\times \dfrac{453.6 \text{ g}}{1 \text{ oz}} \times \dfrac{0.00090 \text{ mol Na}}{144.1}(6.022 \times 10^{23})$

   = 3.2×10^{20} Na atoms

79.   750.0 mL $\times$ 0.110 $\times$ 0.789 g/mL $C_2H_5OH \times \dfrac{180.2 \text{ g C}_6H_{12}O_6}{92.14 \text{ g C}_2H_5OH}$

   = 127 g glucose

81. $24.30 \text{ g/mol} = 0.0272 \mathcal{M}; \mathcal{M} = 893 \text{ g/mol}$

82. $107.9 \text{ g} \times \dfrac{1 \text{ cm}^3}{10.5 \text{ g}} \times \dfrac{10^{21} \text{ nm}^3}{1 \text{ cm}^3} \times \dfrac{4 \text{ atoms}}{(0.409 \text{ nm})^3} = 6.01 \times 10^{23} \text{ atoms}$

83. $\text{mass CaO} = 4.832 \text{ g Ca(OH)}_2 \times \dfrac{1 \text{ mol Ca(OH)}_2}{74.10 \text{ g Ca(OH)}_2} \times \dfrac{1 \text{ mol CaO}}{1 \text{ mol Ca(OH)}_2}$

$\times \dfrac{56.08 \text{ g CaO}}{1 \text{ mol CaO}} = 3.657 \text{ g CaO}$

$\text{mass Ca to CaO} = 3.657 \text{ g CaO} \times \dfrac{40.08 \text{ g Ca}}{56.08 \text{ g CaO}} = 2.614 \text{ g Ca}$

$\text{mass Ca}_3\text{N}_2 = 2.411 \text{ g Ca} \times \dfrac{148.26 \text{ g Ca}_3\text{N}_2}{120.24 \text{ g Ca}} = 2.973 \text{ g Ca}_3\text{N}_2$

84. Let x = mass KBr

$3.595 \text{ g} - x + \dfrac{x(74.55)}{119.00} = 3.129 \text{ g}; \quad x = 1.25 \text{ g}$

$\% \text{ KBr} = \dfrac{1.25 \text{ g}}{3.595 \text{ g}} \times 100\% = 34.7\%$

85. a. 1st oxide: $2.573 \text{ g V} + 2.016 \text{ g O}$

$0.05051 \text{ mol V}, \ 0.1260 \text{ mol O}$

$0.1260/0.05051 = 2.50; \ \text{V}_2\text{O}_5$

2nd oxide: $2.573 \text{ g V} + 1.209 \text{ g O}$

$0.05051 \text{ mol V}, \ 0.07556 \text{ mol O}$

$0.07556/0.05051 = 1.50; \ \text{V}_2\text{O}_3$

b. $2.016 \text{ g O} \times \dfrac{18.02 \text{ g H}_2\text{O}}{16.00 \text{ g O}} = 2.271 \text{ g H}_2\text{O}$

86. $\text{C}_{17}\text{H}_{21}\text{O}_4\text{N} \longrightarrow 17 \text{ CO}_2 \qquad \text{C}_{12}\text{H}_{22}\text{O}_{11} \longrightarrow 12 \text{ CO}_2$

Let x = mass cocaine

$\dfrac{x(748.17)}{(303.35)} \times \dfrac{1}{1.80} + \dfrac{(1.00 - x)(528.12)}{(342.30)} \times \dfrac{1}{1.80} = 1.00$

$1.37x + 0.857 - 0.857x = 1.00$

$x = 0.28; \ 28\%$

CHAPTER 4
Reactions in Aqueous Solution

LECTURE NOTES

This chapter deals with three types of reactions that students will meet with again, in lecture and laboratory: precipitation, acid-base, and oxidation-reduction. Emphasis is placed on the net ionic equations written to represent these reactions and the mass-mole relationships derived from these equations. Students have a lot of trouble writing equations from scratch; balancing redox equations is relatively easy because they follow a set of rules.

Allow three lectures for this chapter. The first deals with writing net ionic equations for precipitation and acid-base reactions. The second lecture covers oxidation number and the balancing of redox equations. The final lecture covers the concept of molarity, solution stoichiometry, and volumetric analysis.

Note that:

1. Students must learn the solubility rules to write equations for precipitation reactions. By the same token, they must know the strong acids and strong bases if they are to write equations for acid-base reactions.

2. Only molecular weak acids and weak bases are considered here; acidic and basic ions are covered in Chapter 13.

3. The half-equation method is the only one described for balancing redox equations. It has the advantage that it gets students into the habit of breaking down the reaction into an oxidation and a reduction. This will come in handy in Chapter 18.

The method used to balance half-equations was suggested to us by Professor Donald Kleinfelter of the University of Tennessee. We prefer it to the conventional procedure, particularly in basic solution, where OH^- ions are introduced directly.

LECTURE 1

I <u>Precipitation Reactions</u>

A. Note that ionic compounds break up in water to form cations and anions.

$$NaOH(s) \longrightarrow Na^+(aq) + OH^-(aq)$$

$$K_2CrO_4(s) \longrightarrow 2K^+(aq) + CrO_4^{2-}(aq)$$

B. <u>Solubility rules</u> (Table 4.1)
Use in predicting results of precipitation reactions.

   1. Mix solutions of $Ba(NO_3)_2$ and Na_2CO_3. What happens?

      Ions present: Ba^{2+}, NO_3^-, Na^+, CO_3^{2-}

      Possible precipitates: $BaCO_3$, $NaNO_3$

      According to solubility rules, $BaCO_3$ is insoluble:

$$Ba^{2+}(aq) + CO_3^{2-}(aq) \longrightarrow BaCO_3(s)$$

   2. Mix solutions of $BaCl_2$, NaOH

      Ions present: Ba^{2+}, Cl^-, Na^+, OH^-

      Possible precipitates: $Ba(OH)_2$, $NaCl$

      Both are soluble; no reaction

II Acids and Bases

A. <u>Acid</u> forms H^+ ion in water

Strong acids are completely ionized in water (HCl, HBr, HI, HNO_3, $HClO_4$, H_2SO_4).

$$HCl(aq) \longrightarrow H^+(aq) + Cl^-(aq)$$

Reaction goes to completion; no HCl molecules in solution

Weak acids are partially ionized in water

$$HF(aq) \rightleftharpoons H^+(aq) + F^-(aq)$$

Hydrofluoric acid contains a mixture of HF molecules, H^+ and F^- ions.

B. <u>Base</u> forms OH^- ions in solution

Strong base is completely ionized in water (Group 1 and heavier Group 2 hydroxides).

$$Ca(OH)_2(s) \longrightarrow Ca^{2+}(aq) + 2\,OH^-(aq)$$

Weak base is partially ionized to form OH^- ions

$$NH_3(aq) + H_2O \rightleftharpoons NH_4^+(aq) + OH^-(aq)$$

C. Acid-base reactions

    1. Strong acid + strong base: $HCl + Ca(OH)_2$

$$H^+(aq) + OH^-(aq) \longrightarrow H_2O$$

    2. Strong base + weak acid: $HF + Ca(OH)_2$

$$HF(aq) + OH^-(aq) \longrightarrow F^-(aq) + H_2O$$

    3. Strong acid + weak base: HCl, NH_3

$$H^+(aq) + NH_3(aq) \longrightarrow NH_4^+(aq)$$

LECTURE 2

I Redox Reactions

  A. Oxidation number; "pseudocharge" assigned according to arbitrary
rules.

    1. Oxidation number of element in elementary substance (e.g., F_2,
O_2) is zero.

    2. Oxidation number of element in monatomic ion is the charge of that
ion. Oxidation number of iron is +2 in Fe^{2+}, +3 in Fe^{3+}.

    3. Oxidation number of Group 1 elements in their compounds is +1;
+2 for Group 2; −1 for fluorine

Oxid. no. H almost always +1; Oxid. no. O ordinarily −2

    4. Sum of oxidation numbers of all atoms in molecule = 0. In poly-
atomic ion, the sum is the charge of the ion.

H_2SO_4: +2 + (oxid. no. S) + 4(−2) = 0; oxid no S = +6

$Cr_2O_7^{2-}$: 2(oxid. no. Cr) + 7(−2) = −2; oxid. no. Cr = +6

  B. Oxidation = increase in oxidation number; reduction = decrease in
oxidation number.

$$HCl(g) + HNO_3(1) \longrightarrow NO_2(g) + \tfrac{1}{2}Cl_2(g) + H_2O(1)$$

HCl is oxidized; oxid. no. Cl increases from −1 to 0

HNO_3 is reduced; oxid. no. N decreases from +5 to +4

II Balancing Redox Equations

  A. $ClO_3^-(aq) + I^-(aq) \longrightarrow Cl^-(aq) + I_2(s)$

1. Split into two half-equations

$$\text{oxidation: } I^-(aq) \;\rightarrow\; I_2(s)$$

$$\text{reduction: } ClO_3^-(aq) \;\rightarrow\; Cl^-(aq)$$

2. Balance half-equations separately, in the following order

   a. Balance atoms of element being oxidized or reduced

$$2I^-(aq) \;\rightarrow\; I_2(s)$$

   b. Balance oxidation number by adding electrons

$$2I^-(aq) \;\rightarrow\; I_2(s) + 2e^-$$

$$ClO_3^-(aq) + 6e^- \;\rightarrow\; Cl^-(aq)$$

   c. Balance charge by adding H^+(acid) or OH^- (base)

$$\text{acid: } ClO_3^-(aq) + 6H^+(aq) + 6e^- \;\rightarrow\; Cl^-(aq)$$

$$\text{base: } ClO_3^-(aq) + 6e^- \;\rightarrow\; Cl^-(aq) + 6\,OH^-(aq)$$

   d. Balance oxygen by adding H_2O molecules

$$\text{acid: } ClO_3^-(aq) + 6H^+(aq) + 6e^- \;\rightarrow\; Cl^-(aq) + 3H_2O$$

$$\text{base: } ClO_3^-(aq) + 3H_2O + 6e^- \;\rightarrow\; Cl^-(aq) + 6\,OH^-(aq)$$

3. Combine half-equations so that electrons cancel

$$\text{acid: } 3[2I^-(aq) \;\rightarrow\; I_2(s) + 2e^-]$$

$$ClO_3^-(aq) + 6H^+(aq) + 6e^- \;\rightarrow\; Cl^-(aq) + 3H_2O$$

$$\overline{6I^-(aq) + ClO_3^-(aq) + 6H^+(aq) \;\rightarrow\; 3I_2(s) + Cl^-(aq) + 3H_2O}$$

$$\text{base: } 6I^-(aq) + ClO_3^-(aq) + 3H_2O \;\rightarrow\; 3I_2(s) + Cl^-(aq) + 6\,OH^-(aq)$$

<u>LECTURE 3</u>

I <u>Molarity</u>

The amount of solute in a solution sample depends upon two factors: the volume of solution and the concentration of solute.

$$\text{Molarity (M)} = \frac{\text{no. moles solute}}{\text{no. liters solution}}$$

1. What volume of 12 M HCl must be taken to obtain 0.10 mol of HCl?

$$\text{Volume} = 0.10 \text{ mol HCl} \times \frac{1 \text{ L}}{12 \text{ mol HCl}} = 0.0083 \text{ L} = 8.3 \text{ mL}$$

2. What mass of NaOH is contained in 125 mL of 6.00 M NaOH?

$$125 \text{ mL} \times \frac{1 \text{ L}}{1000 \text{ mL}} \times \frac{6.00 \text{ mol}}{1 \text{ L}} \times \frac{40.00 \text{ g}}{1 \text{ mol}} = 30.0 \text{ g NaOH}$$

3. What are the molarities of Al^{3+} and SO_4^{2-} in 0.100 M $Al_2(SO_4)_3$?

$$Al_2(SO_4)_3(s) \longrightarrow 2Al^{3+}(aq) + 3SO_4^{2-}(aq)$$

$$[Al^{3+}] = 0.200 \text{ M} \qquad [SO_4^{2-}] = 0.300 \text{ M}$$

II <u>Solution Stoichiometry</u>

A. $Cu^{2+}(aq) + 2\,OH^-(aq) \longrightarrow Cu(OH)_2(s)$

What volume of 0.200 M $CuSO_4$ solution is required to react with 50.0 mL of 0.100 M NaOH?

$$n\, OH^- = 0.0500 \text{ L} \times \frac{0.100 \text{ mol NaOH}}{1 \text{ L}} \times \frac{1 \text{ mol } OH^-}{1 \text{ mol NaOH}} = 0.00500 \text{ mol } OH^-$$

$$n\, Cu^{2+} = 0.00500 \text{ mol } OH^- \times \frac{1 \text{ mol } Cu^{2+}}{2 \text{ mol } OH^-} = 0.00250 \text{ mol } Cu^{2+}$$

$$V = 0.00250 \text{ mol } Cu^{2+} \times \frac{1 \text{ mol } CuSO_4}{1 \text{ mol } Cu^{2+}} \times \frac{1 \text{ L}}{0.200 \text{ mol}} = 0.0125 \text{ L} = 12.5 \text{ mL}$$

B. $6I^-(aq) + ClO_3^-(aq) + 6H^+(aq) \longrightarrow 3I_2(s) + Cl^-(aq) + 3H_2O$

Suppose 22.0 mL of 0.150 M $KClO_3$ is required to react with a sample weighing 5.00 g. Percent of I^- in sample?

$$n\, ClO_3^- = 0.0220 \text{ L} \times \frac{0.150 \text{ mol } ClO_3^-}{1 \text{ L}} = 0.00330 \text{ mol } ClO_3^-$$

$$\text{mass } I^- = 0.00330 \text{ mol } ClO_3^- \times \frac{6 \text{ mol } I^-}{1 \text{ mol } ClO_3^-} \times \frac{126.9 \text{ g } I^-}{1 \text{ mol } I^-} = 2.51 \text{ g } I^-$$

$$\%\, I^- = \frac{2.51 \text{ g}}{5.00 \text{ g}} \times 100 = 50.2$$

34

DEMONSTRATIONS

1. Ionization in water solution: GILB E 2, E 3

2. Conductivity of water solutions: SHAK <u>3</u> 140

3. "Insolubility" of barium nitrate: GILB F 32

4. Solubility of carbonates: GILB M 98

5. Precipitation of lead iodide: GILB M 129

6. Classical properties of acids and bases: SHAK <u>3</u> 58

7. Strength of acids: SHAK <u>3</u> 136

8. Antacids: SHAK <u>3</u> 162

9. Reaction of barium hydroxide with sulfuric acid: GILB   33, E 13

10. Silver and lead trees: GILB A 27

11. Oxidation by sodium peroxide: GILB G 2; SHAK <u>1</u> 59

12. Oxidation by potassium permanganate: GILB G 9

13. Redox reaction of phosphorus with copper(II): GILB G 13

14. Beating heart (redox reaction): GILB G 14

15. Breathalyzer: J. Chem. Educ. <u>67</u> 263 (1990); <u>71</u> 158 (1994)

VIDEO

1. Electrical conductivity: JCES 7

2. Reaction of zinc with sulfur: SHAK 23

3. Oxidation with potassium permanganate: JCES 14

4. Beating heart: Falcon 23

5. Simple titration: JCES 12

PROBLEMS

1. a. $AlCl_3$(soluble)   b. $(NH_4)_2CO_3$ (soluble)   c. $SrSO_4$(insoluble)
   d. $Cr(OH)_3$(insoluble)

3. a. sodium sulfide   b. sodium carbonate   c. sodium hydroxide

5. a. $Fe^{3+}(aq) + 3\ OH^-(aq) \longrightarrow Fe(OH)_3(s)$

   b. $Cd^{2+}(aq) + S^{2-}(aq) \longrightarrow CdS(s)$

    $Sr^{2+}(aq) + SO_4{}^{2-}(aq) \longrightarrow SrSO_4(s)$

7. a. no reaction   b. $Hg_2{}^{2+}(aq) + 2Cl^-(aq) \longrightarrow Hg_2Cl_2(s)$

   c. no reaction   d. $2Ag^+(aq) + CO_3{}^{2-}(aq) \longrightarrow Ag_2CO_3(s)$

   e. $Al^{3+}(aq) + 3\ OH^-(aq) \longrightarrow Al(OH)_3(s)$

    $Sr^{2+}(aq) + SO_4{}^{2-}(aq) \longrightarrow SrSO_4(s)$

9. a. $Sr^{2+}(aq) + CO_3{}^{2-}(aq) \longrightarrow SrCO_3(s)$

   b. $Zn^{2+}(aq) + S^{2-}(aq) \longrightarrow ZnS(s)$

   c. $Hg_2{}^{2+}(aq) + 2Cl^-(aq) \longrightarrow Hg_2Cl_2(s)$

   d. no reaction

11. a. no reaction   b. 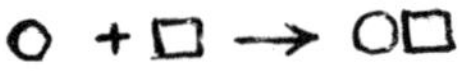 ○ + □ → ○□

13. a. 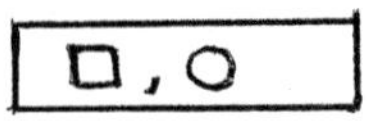 | □ , ○ | b.  | □○ , □○ |

   c.  | □ , ○ | d.

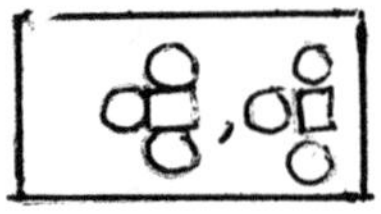

15. a. HNO_2 b. H^+ c. H^+ c. H^+ d. $HClO_2$ e. $HCHO_2$

17. a. OH^- b. NH_3 c. $(CH_3)_3N$ d. OH^-

19. a. strong acid   b. strong base   c. weak acid   d. weak base

21. a. $HC_4H_7O_2(aq) + OH^-(aq) \longrightarrow C_4H_7O_2{}^-(aq) + H_2O$

   b. $C_3H_8NH(aq) + H^+(aq) \longrightarrow C_3H_8NH_2{}^+(aq)$

   c. $HCN(aq) + OH^-(aq) \longrightarrow CN^-(aq) + H_2O$

23. a. $C_2H_5NH_2(aq) + H^+(aq) \longrightarrow C_2H_5NH_3{}^+(aq)$

   b. $H^+(aq) + OH^-(aq) \longrightarrow H_2O$

   c. $HC_2H_3O_2(aq) + OH^-(aq) \longrightarrow C_2H_3O_2{}^-(aq) + H_2O$

d. $H^+(aq) + OH^-(aq) \longrightarrow H_2O$

e. $H^+(aq) + OH^-(aq) \longrightarrow H_2O$

25. a. NO_2; $N = +4$, $O = -2$ b. SO_3; $S = +6$, $O = -2$

    c. MnO_4^-; $Mn = +7$, $O = -2$ d. ClO_3^-; $Cl = +5$, $O = -2$

27. a. $S = +6$, $F = -1$ b. $Sb = +5$, $O = -2$ c. $N = -2$, $H = +1$

    d. $S = +2$, $O = -2$ e. $Cr = +3$, $O = -2$, $H = +1$

29. a. O   b. R   c. O   d. O

31. a. $Ti^{3+}(aq) + 2H_2O \longrightarrow TiO_2(s) + e^- + 4H^+(aq)$

    b. $Sn^{4+}(aq) + 2e^- \longrightarrow Sn^{2+}(aq)$

    c. $H_2O_2(aq) + 2\ OH^-(aq) \longrightarrow O_2(g) + 2e^- + 2H_2O$

    d. $CH_3OH(aq) + 2\ OH^-(aq) \longrightarrow CH_2O(aq) + 2e^- + 2H_2O$

33. a. O: $Mn^{2+}(aq) + 4H_2O \longrightarrow MnO_4^-(aq) + 8H^+(aq) + 5e^-$

    b. R: $CrO_4^{2-}(aq) + 3e^- + 4H_2O \longrightarrow Cr(OH)_4^-(aq) + 4\ OH^-(aq)$

    c. O: $Bi^{3+}(aq) + 6\ OH^-(aq) \longrightarrow BiO_3^-(aq) + 2e^- + 3H_2O$

    d. R: $VO^{2+}(aq) + e^- + 2H^+(aq) \longrightarrow V^{3+}(aq) + H_2O$

35. a. $H_2O_2(aq) \longrightarrow H_2O$; H_2O_2 reduced, oxidizing agent

       $Fe^{2+}(aq) \longrightarrow Fe^{3+}(aq)$; Fe^{2+} oxidized, reducing agent

    b. $C_2H_4(g) \longrightarrow CO_2(g)$; C_2H_4 oxidized, reducing agent

       $O_2(g) \longrightarrow H_2O$; O_2 reduced, oxidizing agent

37. a. $H_2O_2(aq) + 2e^- + 2H^+(aq) \longrightarrow 2H_2O$

$$\underline{2Fe^{2+}(aq) \longrightarrow 2Fe^{3+}(aq) + 2e^-}$$

$$H_2O_2(aq) + 2H^+(aq) + 2Fe^{2+}(aq) \longrightarrow 2H_2O + 2Fe^{3+}(aq)$$

    b. $C_2H_4(g) + 4H_2O \longrightarrow 2CO_2(g) + 12e^- + 12H^+(aq)$

$$\underline{O_2(g) + 4e^- + 4H^+(aq) \longrightarrow 2H_2O}$$

$$C_2H_4(g) + 3\ O_2(g) \longrightarrow 2CO_2(g) + 2H_2O$$

39. a. $Fe^{2+}(aq) \longrightarrow Fe^{3+}(aq) + e^-$

$$\underline{IO_4^-(aq) + 8e^- + 8H^+(aq) \longrightarrow I^-(aq) + 4H_2O}$$

$8Fe^{2+}(aq) + IO_4^-(aq) + 8H^+(aq) \longrightarrow 8Fe^{3+}(aq) + I^-(aq) + 4H_2O$

b. $PbO_2(s) + 2e^- + 2Br^-(aq) + 4H^+(aq) \longrightarrow PbBr_2(s) + 2H_2O$

$$\underline{2H_2O \longrightarrow O_2(g) + 4H^+(aq) + 4e^-}$$

$2PbO_2(s) + 4Br^-(aq) + 4H^+(aq) \longrightarrow 2PbBr_2(s) + 2H_2O + O_2(g)$

c. $Ca(s) \longrightarrow Ca^{2+}(aq) + 2e^-$

$$\underline{VO_4^{3-}(aq) + 3e^- + 8H^+(aq) \longrightarrow V^{2+}(aq) + 4H_2O}$$

$3Ca(s) + 2VO_4^{3-}(aq) + 16H^+(aq) \longrightarrow 3Ca^{2+}(aq) + 2V^{2+}(aq) + 8H_2O$

d. $3IO_3^-(aq) + 18H^+(aq) + 16e^- \longrightarrow I_3^-(aq) + 9H_2O$

$$\underline{3I^-(aq) \longrightarrow I_3^-(aq) + 2e^-}$$

$3IO_3^-(aq) + 18H^+(aq) + 24I^-(aq) \longrightarrow 9I_3^-(aq) + 9H_2O$

$IO_3^-(aq) + 6H^+(aq) + 8I^-(aq) \longrightarrow 3I_3^-(aq) + 3H_2O$

41. a. $Ni(OH)_2(s) + 2e^- \longrightarrow Ni(s) + 2\ OH^-(aq)$

$$\underline{N_2H_4(aq) + 4\ OH^-(aq) \longrightarrow N_2(g) + 4e^- + 4H_2O}$$

$2Ni(OH)_2(s) + N_2H_4(aq) \longrightarrow 2Ni(s) + 4H_2O + N_2(g)$

b. $Fe(OH)_3(s) + e^- \longrightarrow Fe(OH)_2(s) + OH^-(aq)$

$$\underline{Cr(OH)_4^-(aq) + 4\ OH^-(aq) \longrightarrow CrO_4^{2-}(aq) + 3e^- + 4H_2O}$$

$3Fe(OH)_3(s) + Cr(OH)_4^-(aq) + OH^-(aq) \longrightarrow 3Fe(OH)_2(s) + CrO_4^{2-}(aq) + 4H_2O$

c. $MnO_4^-(aq) + 3e^- + 2H_2O \longrightarrow MnO_2(s) + 4\ OH^-(aq)$

$$\underline{BrO_3^-(aq) + 2\ OH^-(aq) \longrightarrow BrO_4^-(aq) + 2e^- + H_2O}$$

$2MnO_4^-(aq) + 3BrO_3^-(aq) + H_2O \longrightarrow 2MnO_2(s) + 2\ OH^-(aq) + 3BrO_4^-(aq)$

d. $H_2O_2(aq) + 2\ OH^-(aq) \longrightarrow O_2(g) + 2H_2O + 2e^-$

$$\underline{IO_4^-(aq) + 4e^- + 2H_2O \longrightarrow IO_2^-(aq) + 4\ OH^-(aq)}$$

$2H_2O_2(aq) + IO_4^-(aq) \longrightarrow 2\ O_2(g) + 2H_2O + IO_2^-(aq)$

43. a. $Ag(s) \longrightarrow Ag^+(aq) + e^-$

$$NO_3^-(aq) + e^- + 2H^+(aq) \longrightarrow NO_2(g) + H_2O$$

$$Ag(s) + 2H^+(aq) + NO_3^-(aq) \longrightarrow Ag^+(aq) + NO_2(g) + H_2O$$

b. $CuS(s) \longrightarrow Cu^{2+}(aq) + S(s) + 2e^-$

$$NO_3^-(aq) + 3e^- + 4H^+(aq) \longrightarrow NO(g) + 2H_2O$$

$$3CuS(s) + 2NO_3^-(aq) + 8H^+(aq) \longrightarrow 3Cu^{2+}(aq) + 3S(s) + 2NO(g) + 4H_2O$$

c. $Sn^{2+}(aq) \longrightarrow Sn^{4+}(aq) + 2e^-$

$$IO_4^-(aq) + 8e^- + 8H^+(aq) \longrightarrow I^-(aq) + 4H_2O$$

$$4Sn^{2+}(aq) + IO_4^-(aq) + 8H^+(aq) \longrightarrow 4Sn^{4+}(aq) + I^-(aq) + 4H_2O$$

45. $W > Z$; $X > W$; $X > Y$; $Z > Y$; hence, $X > W > Z > Y$

47. $0.483 \text{ mol/L} \times 0.615 \text{ L} = 0.297 \text{ mol}$

a. $\mathcal{M} = (39.10 + 54.94 + 64.00)\text{g/mol} = 158.04 \text{ g/mol}$

mass $KMnO_4 = 158.04 \times 0.297 \text{ g} = 46.9 \text{ g}$

dissolve 46.9 g to form 615 mL of solution

b. $\mathcal{M} = (22.99 + 35.45)\text{g/mol} = 58.44 \text{ g/mol}$

mass $NaCl = 58.44 \times 0.297 \text{ g} = 17.4 \text{ g}$

dissolve 17.4 g to form 615 mL of solution

c. $\mathcal{M} = (72.06 + 8.06 + 96.00)\text{g/mol} = 176.12 \text{ g/mol}$

mass $= 176.12 \times 0.297 \text{ g} = 52.3 \text{ g}$

dissolve 52.3 g to form 615 mL of solution

49. a. $0.450 \text{ mol/L} \times 0.0456 \text{ L} = 0.0205 \text{ mol}$

b. $\dfrac{0.800 \text{ mol}}{0.450 \text{ mol/L}} \times 10^3 \text{ mL} = 1.78 \times 10^3 \text{ mL}$

c. need 2.00 mol K_2CO_3; have 0.900 mol

$1.10 \text{ mol} \times 138.2 \text{ g/mol} = 152 \text{ g}$

d. $\dfrac{50.0}{125} \times 0.450 \text{ mol/L} = 0.180 \text{ mol/L}$

51. a. $0.625 \text{ mol } Mg^{2+}$, $1.25 \text{ mol } Cl^-$; total $= 1.88 \text{ mol}$

b. 1.250 mol Al^{3+}, 1.88 mol SO_4^{2-}; total = 3.13 mol

c. 0.625 mol Co^{3+}, 1.88 mol NO_3^{-}; total = 2.50 mol

d. 0.625 mol Fe^{2+}, 1.25 mol MnO_4^{-}; total = 1.88 mol

53. a. $Pb^{2+}(aq) + SO_4^{2-}(aq) \longrightarrow PbSO_4(s)$

$$0.0150 \times 0.0356 \text{ mol } SO_4^{2-} \times \frac{1 \text{ mol } Pb^{2+}}{1 \text{ mol } SO_4^{2-}} \times \frac{1 \text{ L}}{0.250 \text{ mol } Pb^{2+}} = 0.00214 \text{ L}$$

b. $Pb^{2+}(aq) + 2Cl^{-}(aq) \longrightarrow PbCl_2(s)$

$$0.0382 \times 0.458 \text{ mol } Cl^{-} \times \frac{1 \text{ mol } Pb^{2+}}{2 \text{ mol } Cl^{-}} \times \frac{1 \text{ L}}{0.250 \text{ mol } Pb^{2+}} = 0.0350 \text{ L}$$

c. $Pb^{2+}(aq) + 2 OH^{-}(aq) \longrightarrow Pb(OH)_2(s)$

$$0.0350 \times 0.297 \text{ mol } OH^{-} \times \frac{1 \text{ mol } Pb^{2+}}{2 \text{ mol } OH^{-}} \times \frac{1 \text{ L}}{0.250 \text{ mol } Pb^{2+}} = 0.0208 \text{ L}$$

55. a. $Hg_2^{2+}(aq) + 2Cl^{-}(aq) \longrightarrow Hg_2Cl_2(s)$

$n\ Hg_2^{2+} = 0.0250 \times 0.0500$ mol; $n\ Cl^{-} = 0.0250 \times 0.0500 \times 2$ mol

$n\ ScCl_3 = 0.0250 \times 0.0500 \times 2/3$ mol

$$V = \frac{0.0250 \times 0.0500 \times 2 \text{ mol}}{3 \times 0.0500 \text{ mol/L}} = 0.0167 \text{ L}$$

b. $0.0250 \times 0.0500 \text{ mol} \times \dfrac{472.1 \text{ g}}{1 \text{ mol}} = 0.590 \text{ g}$

57. $n\ Ba(OH)_2 = 0.216/171.3$; $n\ HNO_3 = 0.216 \times 2/171.3$

$$M = \frac{0.216 \times 2 \text{ mol}}{171.3 \times 0.0200 \text{ L}} = 0.126 \text{ mol/L}$$

59. a. $H^{+}(aq) + OH^{-}(aq) \longrightarrow H_2O$

$$\frac{0.02500 \times 0.418 \text{ mol}}{0.550 \text{ mol/L}} = 0.0190 \text{ L}$$

b. $H^{+}(aq) + OH^{-}(aq) \longrightarrow H_2O$

$$\frac{10.00 \times 2 \text{ mol}}{74.09} \times \frac{1 \text{ L}}{0.550 \text{ mol}} = 0.491 \text{ L}$$

c. $\dfrac{15.0 \times 0.928 \times 0.100}{31.06}$ mol $\times$ $\dfrac{1\ L}{0.550\ mol}$ $= 0.0815\ L$

61. a. $MnO_4^-(aq) + 3e^- + 4H^+(aq) \longrightarrow MnO_2(s) + 2H_2O$

$H_2C_2O_4(aq) \longrightarrow 2CO_2(g) + 2e^- + 2H^+(aq)$

$\overline{2MnO_4^-(aq) + 3H_2C_2O_4(aq) + 2H^+(aq) \longrightarrow 2MnO_2(s) + 4H_2O + 6CO_2(g)}$

b. $n\ MnO_4^- = 0.0250 \times 0.500$ mol

$n\ C_2O_4^{2-} = 0.0250 \times 0.500 \times 3/2$ mol

$M = \dfrac{0.0250 \times 0.500 \times 3/2\ mol}{0.0150\ L} = 1.25\ mol/L$

c. 0.0125 mol $\times 86.94$ g/mol $= 1.09$ g

63. a. $Ag(s) + 2H^+(aq) + NO_3^-(aq) \longrightarrow Ag^+(aq) + NO_2(g) + H_2O$

b. 0.03500×12.0 mol $H^+ \times \dfrac{1\ mol\ Ag^+}{2\ mol\ H^+} \times \dfrac{107.9\ g\ Ag}{1\ mol\ Ag} = 22.7\ g\ Ag$

65. mass $NaHCO_3 = 0.100 \times 3.00$ mol $\times 84.01$ g/mol $= 25.2$ g $NaHCO_3$

67. mass $NaHCO_3 = 0.215 \times 0.0187 \times 84.01$ g

% $NaHCO_3 = \dfrac{0.215 \times 0.0187 \times 84.01}{0.400} \times 100 = 84.4\%$

69. n lactic acid $= \dfrac{0.100\ g}{90.08\ g/mol} = 1.11 \times 10^{-3}$ mol

$n\ OH^- = 0.01295 \times 0.0857 = 1.11 \times 10^{-3}$ mol

1 mol

71. $H_2O_2(aq) \longrightarrow O_2(g) + 2e^- + 2H^+(aq)$

$Cr_2O_7^{2-}(aq) + 6e^- + 14H^+(aq) \longrightarrow 2Cr^{3+}(aq) + 7H_2O$

$\overline{Cr_2O_7^{2-}(aq) + 8H^+(aq) + 3H_2O_2(aq) \longrightarrow 2Cr^{3+}(aq) + 7H_2O + 3\ O_2(g)}$

$n\ Cr_2O_7^{2-}(aq) = 0.388 \times 0.0758$ mol

mass $H_2O_2 = 0.388 \times 0.0758 \times 3 \times 34.02$ g

% $= \dfrac{0.388 \times 0.0758 \times 3 \times 34.02}{15.0} \times 100 = 20.0\%$

73. n AgCl = n HClO = $\dfrac{3.49}{143.4}$;   mass HClO = $\dfrac{3.49}{143.4}$ x 52.46 g

mass % HClO = $\dfrac{3.49 \text{ x } 52.46}{143.4 \text{ x } 25.0 \text{ x } 1.02}$ x 100 = 5.01 %

75. a. precipitation   b. acid-base   c. redox   d. redox

77. $MnO_4^-(aq) + 8H^+(aq) + 5e^- \longrightarrow Mn^{2+}(aq) + 4H_2O$

$\underline{Fe^{2+}(aq) \longrightarrow Fe^{3+}(aq) + e^-}$

$MnO_4^-(aq) + 8H^+(aq) + 5Fe^{2+}(aq) \longrightarrow Mn^{2+}(aq) + 4H_2O + 5Fe^{3+}(aq)$

n MnO_4^- = 0.0323 x 0.002100; n Fe^{2+} = 0.0323 x 0.002100 x 5

% Fe = $\dfrac{0.0323 \text{ x } 0.002100 \text{ x } 5 \text{ x } 55.85}{5.00}$ x 100 = 0.379%

79. $5CaC_2O_4(s) + 2MnO_4^-(aq) + 16H^+(aq) \longrightarrow 10CO_2(g) + 2Mn^{2+}(aq) + 5Ca^{2+}(aq)$

$+ 8\ H_2O$

n MnO_4^- = 0.0262 L x 0.0946 mol/L = 2.48 x 10^{-3} mol

n CaC_2O_4 = 2.48 x 10^{-3} mol x 5/2 = 6.20 x 10^{-3} mol ; 0.794 g

mass Ca = 6.20 x 10^{-3} mol x 40.08 g/mol = 0.248 g; yes

80. n $Mg(OH)_2$ = 0.330 g x 0.410 x $\dfrac{1 \text{ mol}}{58.32 \text{ g}}$ = 0.00232 mol

n $NaHCO_3$ = 0.330 g x 0.362 x $\dfrac{1 \text{ mol}}{84.01 \text{ g}}$ = 0.00142 mol

n H^+ = 2(0.00232 mol) + 0.00142 mol = 0.00606 mol

V = 0.00606 mol x $\dfrac{1 \text{ L}}{0.020 \text{ mol}}$ = 0.30 L

81. x = mass of Cu reacting; 215.8/63.55 = 3.40x = mass of Ag formed

2.00 - x + 3.40x = 4.18; solving, x = 0.908 g

mass Cu = 1.09 g; mass Ag = 3.09 g

82. $5Fe^{2+}(aq) + MnO_4^-(aq) + 8H^+(aq) \longrightarrow 5Fe^{3+}(aq) + Mn^{2+}(aq) + 4H_2O$

1) n MnO_4^- = 0.0350 L x 0.0280 mol/L = 9.80 x 10^{-4} mol

n Fe^{2+} = 5 x 9.80 x 10^{-4} mol = 4.90 x 10^{-3} mol

$$M\ Fe^{2+} = \frac{4.90 \times 10^{-3}\ mol}{5.000 \times 10^{-2}\ L} = 0.0980\ mol/L$$

2) $n\ MnO_4^{-} = 0.0480\ L \times 0.0280\ mol/L = 1.34 \times 10^{-3}\ mol$

$n\ Fe^{2+} = 5 \times 1.34 \times 10^{-3}\ mol = 6.72 \times 10^{-3}\ mol$

$n\ Fe^{3+} = 6.72 \times 10^{-3}\ mol - 4.90 \times 10^{-3}\ mol = 1.82 \times 10^{-3}\ mol$

$$M\ Fe^{3+} = \frac{1.82 \times 10^{-3}\ mol}{5.00 \times 10^{-2}\ L} = 0.0364\ mol/L$$

CHAPTER 5
Gases

LECTURE NOTES

The ideal gas law is at the heart of this chapter. It is unnecessary to spend much time on the relationships between variables such as Boyle's law or Charles' law. For one thing, students almost certainly got a heavy dose of problems of that type in high school. Moreover, such relationships have very limited application to chemistry.

Dalton's law is another one that students have very little trouble with. The relation between mole fraction and partial pressure, though, is important; it comes up later in the guise of Raoult's law or Henry's law. We ordinarily spend relatively little time on deviations from the ideal gas law. The van der Waals equation is no longer covered; students never used it.

The basic equation of kinetic theory is: $\frac{1}{2}mu^2$ = constant x T. We relate u to T and to molar mass (Graham's law) through this equation. The relation $u = (3RT/M)^{1/2}$ is presented without derivation; be careful about the units for R! The concept of a distribution of molecular veloc- ities is introduced here and serves as background for chemical kinetics.

LECTURE 1

I <u>The Ideal Gas Law</u>

   A. <u>Variables</u> V = volume (liters, cm^3, m^3)
                       n = amount in moles
                       T = temperature (K)
                       P = pressure (atmospheres, mm Hg)

Relation between variables:

$$PV = nRT$$

where R = 0.0821 L•atm/mol•K

Saunders College Publishing

B. <u>Initial and final state problems</u>. Use ideal gas law to find necessary
   relations.

A cylinder contains a gas at a pressure of 255 lb/in^2. If the valve
is opened and 75% of the gas escapes, what is the final pressure?

V and T are constant, so $P_2/P_1 = n_2/n_1$; $n_2 = 0.25n_1$

$P_2 = 0.25P_1 = 0.25 \times 255$ lb/in^2 = 64 lb/in^2

C. <u>Calculation of P, V, n or T</u>

What is the pressure exerted by 15.0 mol of O_2 in a 50.0 L tank at
25° C?

$$P = \frac{nRT}{V} = \frac{(15.0 \text{ mol})(0.0821 \text{ L}\cdot\text{atm/mol}\cdot\text{K})(298 \text{ K})}{50.0 \text{ L}} = 7.34 \text{ atm}$$

Note that since R = 0.0821 L·atm/mol·K, V must be in liters, P in
atmospheres, T in K, n in moles

D. <u>Calculation of density or molar mass</u>

$$PV = \frac{mRT}{\mathcal{M}} \; ; \quad \frac{m}{V} = d = \frac{P\mathcal{M}}{RT}$$

Density of $O_2(g)$ at 27°C, 735 mm Hg?

$$d = \frac{(735/760 \text{ atm})(32.00 \text{ g/mol})}{(0.0821 \text{ L}\cdot\text{atm/mol}\cdot\text{K})(300 \text{ K})} = 1.26 \text{ g/L}$$

A flask weighs 52.693 g empty and 53.117 g when filled with acetone
vapor at 100°C and 752 mm Hg. Taking the volume of the flask to be
226.2 mL, calculate the molar mass of acetone.

$$\mathcal{M} = \frac{mRT}{PV} = \frac{(0.424 \text{ g})(0.0821 \text{ L}\cdot\text{atm/mol}\cdot\text{K})(373 \text{ K})}{(752/760 \text{ atm})(0.2262 \text{ L})} = 58.0 \text{ g/mol}$$

<u>LECTURE 2</u>

I <u>The Ideal Gas Law</u>

A. <u>Volumes of gases involved in reactions</u>

$$Zn(s) + 2H^+(aq) \longrightarrow Zn^{2+}(aq) + H_2(g)$$

Mass of Zn required to form 16.0 L of $H_2(g)$ at 20°C, 735 mm Hg?

Path to follow: V $H_2 \longrightarrow$ n $H_2 \longrightarrow$ n Zn $\longrightarrow$ mass Zn

$$n\ H_2 = \frac{PV}{RT} = \frac{(735/760\ \text{atm})(16.0\ \text{L})}{(0.0821\ \text{L·atm/mol·K})\ (293\ \text{K})} = 0.643\ \text{mol}\ H_2$$

$$\text{mass Zn} = 0.643\ \text{mol}\ H_2 \times \frac{1\ \text{mol Zn}}{1\ \text{mol}\ H_2} \times \frac{65.38\ \text{g Zn}}{1\ \text{mol Zn}} = 42.0\ \text{g Zn}$$

II Dalton's Law (Gas Mixtures)

A. $P_{tot} = P_1 + P_2 + - -$ where P_1 is the partial pressure of gas 1, etc.
Most often used in collection of gases over water:

$$P_{gas} = P_{tot} - P_{H_2O}$$

where P_{H_2O} is the vapor pressure of water.

B. $P_1 = X_1 P_{tot}$

where X = mole fraction of gas in mixture. Partial pressure of oxygen in air (X = 0.2095) when barometric pressure = 734 mm Hg?

$$P_{O_2} = 0.2095(734\ \text{mm Hg}) = 154\ \text{mm Hg}$$

III Kinetic Theory of Gases

$$E_{trans} = \tfrac{1}{2}mu^2 = C \times T$$

where m = mass of molecule, u = average speed, T = temperature in K, and C is a constant which has the same value for all gases.

A. <u>Graham's law</u> Relates m and u for two different gases at same T

$$m_2 u_2^{\ 2} = m_1 u_1^{\ 2}$$

$$(\text{rate 2})/(\text{rate 1}) = (M_1/M_2)^{\frac{1}{2}} = (\text{time 1})/(\text{time 2})$$

Certain gas takes 2.42 times as long to effuse as O_2 at same T, P. Molar mass of gas?

$$(M/32.0\ \text{g/mol})^{\frac{1}{2}} = 2.42; \quad M = (2.42)^2 \times 32.0\ \text{g/mol} = 187\ \text{g/mol}$$

B. $u = (3RT/M)^{\frac{1}{2}}$ where $R = 8.31 \times 10^3$ to get u in meters/second

Average velocity of H_2 molecule at 0°C?

$$u = (3 \times 8.31 \times 10^3 \times 273/2.02)^{\frac{1}{2}} = 1.84 \times 10^3\ \text{m/s}$$

IV Real Gases

Deviate at least slightly from ideal gas law because of two factors:

1. Gas molecules attract each other
2. Gas molecules occupy a finite volume
Both of these factors are neglected in the ideal gas law. Both increase
in importance when the molecules are close together (high P, low T).

DEMONSTRATIONS

1. Boyle's law: GILB D 28, D 30, D 38; SHAK 2 12, 20

2. Charles' law: GILB D 31, D 40; SHAK 2 24, 28; J. Chem. Educ. 71 433
                                                                  (1994)

3. Avogadro's law: GILB D 36

4. Molar volume of nitrogen: J. Chem. Educ. 69 693 (1992)

5. Molar mass of butane: SHAK 2 48

6. Dalton's law: GILB D 34; SHAK 2 41

7. Law of combining volumes: GILB A 19

8. Electrolysis of water: GILB A 5, G 57

9. Kinetic theory simulator: SHAK 2 96

10. Graham's law: GILB D 6, D 8; SHAK 2 59, 69, 73

11. Properties of gases: GILB D 10

VIDEO

1. Atmospheric pressure: JCES 21

2. Gas density: JCES 22

3. Combining volumes: SHAK 30

4. Electrolysis of water: Falcon 22

5. Graham's law: SHAK 32

PROBLEMS

1. $5.000 \text{ gal} \times \dfrac{4 \text{ qt}}{1 \text{ gal}} \times \dfrac{1 \text{ L}}{1.057 \text{ qt}} = 18.92 \text{ L}$

$0.784 \text{ mol} \times \dfrac{44.09 \text{ g}}{1 \text{ mol}} = 34.6 \text{ g}$

$t_{°C} = 36/1.8 = 20°C; \quad T = 293 \text{ K}$

3.   913 mm Hg, 1.20 atm, 1.22×10^2 kPa

633 mm Hg, 0.833 atm, 84.4 kPa

915 mm Hg, 1.20 atm, 122 kPa

5. a.   b. P is 0.65 times as great at $-80°C$

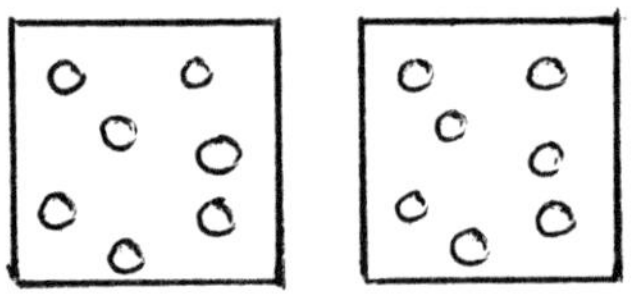

7. a. $1.00 \text{ atm} \times \dfrac{263 \text{ K}}{300 \text{ K}} = 0.877 \text{ atm}$ b. $1.00 \text{ atm} \times \dfrac{313 \text{ K}}{300 \text{ K}} = 1.04 \text{ atm}$

9. a. $V_1/V_2 = 288/303 = 0.950$ b. $V_1/V_2 = 1.50/3.00 = 0.500$

11. $P = 42.7 \text{ psi} \times \dfrac{319 \text{ K}}{295 \text{ K}} = 46.2 \text{ psi}$; gauge P = 31.5 psi

13. $0.500 \text{ L} \times \dfrac{751 \text{ mm Hg}}{760 \text{ mm Hg}} \times \dfrac{310 \text{ K}}{312 \text{ K}} = 0.491 \text{ L}$

15. $1.28 \times 10^3 \text{ L} \times \dfrac{0.998 \text{ atm}}{0.753 \text{ atm}} \times \dfrac{248 \text{ K}}{304 \text{ K}} = 1.38 \times 10^3 \text{ L}$

17. $P = \dfrac{nRT}{V} = \dfrac{0.25 \times 0.0821 \times 310}{32.0} \text{ atm} = 0.20 \text{ atm}$

19. mass He $= 22.2 \text{ g} \times \dfrac{4.00}{29.0} = 3.06 \text{ g}$; mass ball = 503.2 g

21. 34.6 atm, 59.3 g; 1.96 L, 0.06238 g;

0.455 mol, $-149°C$; 0.0695 mol, 1.11 g

23. a. same   b. Tank B greater (larger number of moles)

25. $d = P\mathcal{M}/RT$

a. 1.19 g/L   b. 2.10 g/L   c. 1.90 g/L

27. $d_2/d_1 = \dfrac{75}{1.0} \times \dfrac{298}{733} = 30$

29. a. $\mathcal{M} = dRT/P = \dfrac{4.65 \dfrac{\text{g}}{\text{L}} \times 0.0821 \dfrac{\text{L·atm}}{\text{mol·K}} \times 306 \text{ K}}{735/760 \text{ atm}} = 121 \text{ g/mol}$

b. n C = 121 g $\times$ 0.0992/12.0 = 1; similarly, n Cl = 2, n F = 2

CCl_2F_2

31. a. $0.745(28.01) + 0.157(32.00) + 0.036(44.01) + 0.062(18.02)$

     $= 28.6$ g/mol

  b. $d = \dfrac{28.6 \times 757/760}{0.0821 \times 310} \dfrac{\text{g}}{\text{L}} = 1.12$ g/L vs 1.13 g/L

33. $n = \dfrac{(0.297)(769/760)}{(0.0821)(308)}$ mol $= 0.0119$ mol; $\mathcal{M} = \dfrac{1.58 \text{ g}}{0.0119 \text{ mol}} = 133$ g/mol

  $(133 - 27)/3 = 35.5;$ Cl

35. 3.0 L

37. n HCN $= \dfrac{(0.987)(27 \times 10^3)}{(0.0821)(296)} = 1.1 \times 10^3 =$ n HCl

  $V = \dfrac{1.1 \times 10^3 \text{ mol}}{6.00 \text{ mol/L}} = 1.8 \times 10^2$ L

39. a. $NH_4NO_3(s) \longrightarrow N_2(g) + \tfrac{1}{2} O_2(g) + 2H_2O(g)$

  b. n $NH_4NO_3 = \dfrac{100.0 \text{ g}}{80.05 \text{ g/mol}}$; n gas $= \dfrac{7}{2} \times \dfrac{100.0}{80.05} = 4.37$ mol

    $P = \dfrac{(4.37)(0.0821)(588)}{10.0}$ atm $= 21.1$ atm

41. P $CO_2 = 0.950(735$ mm Hg$) = 698$ mm Hg

  P $O_2 = 0.050(735$ mm Hg$) = 37$ mm Hg

43. P gas $= 1.00$ atm $\times \dfrac{1.04}{1.00} \times \dfrac{315}{363} = 0.902$ atm

  P $H_2O = 0.084$ atm $= 64$ mm Hg

45. a. C   b. 1.00 atm   c. 4.50 atm

  d. 3.50 atm; 4.50 atm   e. 3.00 atm; 4.50 atm

47. $n_A = \dfrac{(2.50)(3.00)}{RT}$; $n_B = \dfrac{(2.00)(0.575)}{RT}$

  $P_{tot} = \dfrac{(2.50)(3.00) + (2.00)(0.575)}{5.00} = 1.73$ atm

49. a. lighter   b. $\dfrac{30.07 \text{ g/mol}}{\mathcal{M}} (1.22)^2; \mathcal{M} = 20.2$ g/mol

51. $u_2/u_1 = (29/4.00)^{\frac{1}{2}} = 2.7$

53. a. $u = \left(\dfrac{3 \times 8.31 \times 10^3 \times 308}{70.90}\right)^{\frac{1}{2}} = 329$ m/s

b. $u = \left(\dfrac{3 \times 8.31 \times 10^3 \times 238}{131.3}\right)^{\frac{1}{2}} = 213$ m/s

55. a. more   b. less

57. a. $V_m = 0.78\ V_m^\circ = 0.78\ RT/P$

$d = \dfrac{16.03\ P}{0.78\ RT} = \dfrac{(16.03)(100)}{(0.78)(0.0821)(298)}$ g/L $= 84$ g/L

b. $d = \dfrac{(16.03)(100)}{(0.0821)(298)}$ g/L $= 65.5$ g/L

c. about 340 atm

59. a. $C_8H_{18}(1) + 25/2\ O_2(g) \longrightarrow 8CO_2(g) + 9H_2O(1)$

b. $\dfrac{50.0\ gal}{26} \times \dfrac{4\ qt}{1\ gal} \times \dfrac{1\ L}{1.057\ qt} \times \dfrac{692g}{1\ L} \times \dfrac{1\ mol}{114.2\ g}$

$= 44.1$ mol $C_8H_{18} = 353$ mol CO_2

$V = \dfrac{(353\ mol)(0.0821\ L\cdot atm/mol\cdot K)(298\ K)}{1.00\ atm} = 8.6 \times 10^3$ L

61. a.

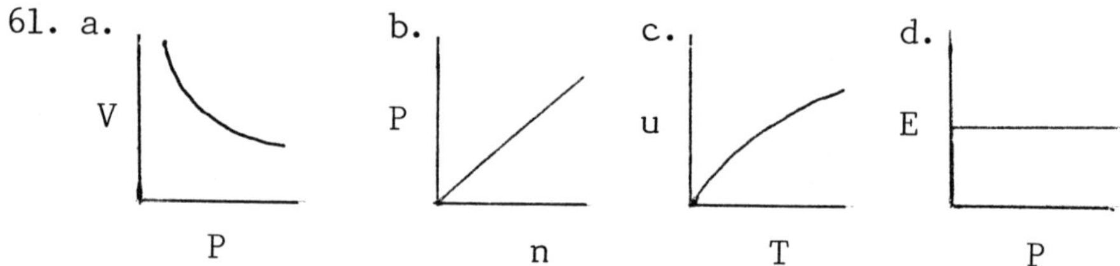

63. a. same   b. Ne   c. Ne   d. Ne

65. a $n\ O_2 = \dfrac{(1.00)(0.618)}{(0.0821)(348)} \times 0.210$ mol $= 4.54 \times 10^{-3}$ mol

b. mass gasoline $= \dfrac{4.54 \times 10^{-3}}{12} \times 100$ g $= 0.038$ g

67. mass $CO_2 = \mathcal{M}\dfrac{PV}{RT} = \dfrac{44.01 \times 1.00 \times 0.1329}{0.0821 \times 298}$ g $= 0.239$ g

% C $= \dfrac{0.239}{0.2036} \times \dfrac{12.01}{44.01} \times 100 = 32.0$; % H $= \dfrac{0.122}{0.2036} \times \dfrac{2.016}{18.02} \times 100 = 6.70$

mass $N_2 = \dfrac{28.01 \times 1.00 \times 0.0408}{0.0821 \times 298} = 0.0467$ g

$\% \text{ N} = \dfrac{0.0467 \text{ g}}{0.2500 \text{ g}} \times 100 = 18.7$

$\% \text{ O} = 100.0 - 32.0 - 6.7 - 18.7 = 42.6$

in 100 g there are 2.66 mol C, 6.65 mol H, 1.33 mol N, 2.66 mol O

$$C_2H_5NO_2$$

69. a. cylinder II; contains more moles

b. compare values of P/T: 7.0/323 > 5.5/283; cylinder I

70. d NH_3/d HCl = $(36.45/17.03)^{\frac{1}{2}} = 1.463$

2.463 d HCl = 5.0 ft; d HCl = 2.0 ft; 3.0 ft from NH_3 end

71. $1.8°R = K$

$0.0821 \dfrac{L \cdot atm}{mol \cdot K} \times \dfrac{1 \text{ K}}{1.8°R} = 0.0456 \dfrac{L \cdot atm}{mol \cdot K}$

72. n H_2 = $\dfrac{(0.147 \text{ L})(755/760 \text{ atm})}{(0.0821 \text{ L} \cdot atm/mol \cdot K)(298 \text{ K})} = 5.97 \times 10^{-3}$ mol

let x = mass Al

$0.00597 = \dfrac{3x/2}{26.98} + \dfrac{0.2500 - x}{65.39}$; x = 0.0533 g;

mass Zn = 0.197 g; % Zn = 0.197/0.250 × 100 = 78.8

73. m air = $\dfrac{(758/760 \text{ atm})(V)(29.0 \text{ g/mol})}{(0.0821 \text{ L} \cdot atm/mol \cdot K)(295 \text{ K})} = 1.19 \text{ V}$

m H_2 = $\dfrac{(758/760 \text{ atm})(V)(2.016 \text{ g/mol})}{(0.0821 \text{ L} \cdot atm/mol \cdot K)(295 \text{ K})} = 0.0830 \text{ V}$

$1.19V = 1.68 \times 10^5 + 0.08V$; $V = 1.51 \times 10^5$ L = 151 m^3

151 m^3 = $4\pi r^3/3$; r = 3.30 m; diameter = 6.60 m

74. $P_2 = P_1 \times \dfrac{T_2}{T_1} \times \dfrac{n_2}{n_1}$; n orig = 3.00; n final = 1.76 + 0.12 + 0.24

= 2.12 mol

P_2 = 0.950 atm × $\dfrac{398}{298}$ × $\dfrac{2.12}{3.00}$ = 0.897 atm

77. $V_a = n_a RT/P$; $V_{tot} = n_{tot} RT/P$

$$V_a/V = n_a/n_{tot} = X_a$$

$$n_a/n_{tot} \neq m_a/m_{tot} \qquad \text{because molar masses differ}$$

Saunders College Publishing

$$n_a/n_{tot} \neq m_a/m_{tot} \qquad \text{because molar masses differ}$$

CHAPTER 6
Electronic Structure and the Periodic Table

LECTURE NOTES

This chapter requires $2\frac{1}{2}$ - 3 lectures. Experience suggests that all four quantum numbers are difficult for students to digest at a single sitting. We prefer to break them up between two lectures.

The "aufbau" approach can be used to predict electronic structures of atoms through atomic number 36. Beyond that, it's probably simpler to use the periodic table to derive electronic structures. Orbital diagrams follow from electron configurations by applying Hund's rule.

In discussing the electronic structure of monatomic ions, it is important to stress the difference between transition metal atoms and cations. The chapter concludes with a description of trends in the periodic table as regards atomic properties.

LECTURE 1

I <u>Atomic Spectra</u>

Produced when electron moves from higher to lower energy level, giving off light in the process

$$\Delta E = E_{hi} - E_{lo} = h\nu = hc/\lambda$$

For the yellow line in the sodium spectrum, $\lambda = 589.0$ nm

$$\nu = \frac{c}{\lambda} = \frac{2.998 \times 10^8 \text{ m/s}}{589.0 \times 10^{-9} \text{ m}} = 5.090 \times 10^{14}/\text{s}$$

$$\Delta E = \frac{(6.626 \times 10^{-34} \text{ J·s})(2.998 \times 10^8 \text{ m/s})}{589.0 \times 10^{-9} \text{ m}} = 3.373 \times 10^{-19} \text{ J}$$

For one mole of electrons:

$$\Delta E = 3.373 \times 10^{-19} \text{ J} \times \frac{6.022 \times 10^{23}}{1 \text{ mol}} \times \frac{1 \text{ kJ}}{10^3 \text{ J}} = 203.1 \text{ kJ}$$

Hence two energy levels in the Na atom differ in energy by 203.1 kJ/mol

II <u>Hydrogen Atom</u>

    A. <u>Bohr model</u> Bohr postulated that electron moves about nucleus in circular orbit of fixed radius. By absorbing energy, it moves to a higher orbit of greater energy; energy given off as electron returns.

$$E_n = (-2.180 \times 10^{-18} \text{ J})/n^2 \qquad n = 1, 2, 3, - -$$

When electron moves from n = 3 to n = 2:

$$E_3 = -2.422 \times 10^{-19} \text{ J}; \quad E_2 = -5.450 \times 10^{-19} \text{ J}$$

$$E_{hi} - E_{lo} = 3.028 \times 10^{-19} \text{ J}$$

$$\lambda = \frac{hc}{\Delta E} = \frac{(6.626 \times 10^{-34} \text{ J} \cdot \text{s})(2.998 \times 10^8 \text{ m/s})}{3.028 \times 10^{-19} \text{ J}}$$

$$= 6.560 \times 10^{-7} \text{ m} = 656.0 \text{ nm (1st line in Balmer series)}$$

    B. <u>Quantum mechanical model</u>
       1. Can only refer to the probability of finding an electron in a region; cannot specify path.
       2. Four quantum numbers required to describe completely the energy of electron in all atoms.

III <u>Electronic Structure</u>

    A. <u>Principal energy levels</u>

n = 1, 2, 3, - -. Value of n is the main factor that determines the energy of an electron and its distance from the nucleus. Maximum capacity of principal level = $2n^2$.

n	1	2	3	4
max. no. e^-	2	8	18	32

    B. <u>Sublevels</u>

      1. Quantum number ℓ = 0, 1, 2, - - (n - 1)

n = 1 ℓ = 0 (one sublevel)
n = 2 ℓ = 0, 1 (two sublevels)
n = 3 ℓ = 0, 1, 2 (three sublevels)

In general, no. of sublevels = n

      2. Sublevel designations: s, p, d, f

value of ℓ	0	1	2	3
letter	s	p	d	f
capacity	2	6	10	14

LECTURE 2

I Electronic Structure

A. <u>Electron configuration</u> Indicate by a superscript the number of electrons in each sublevel

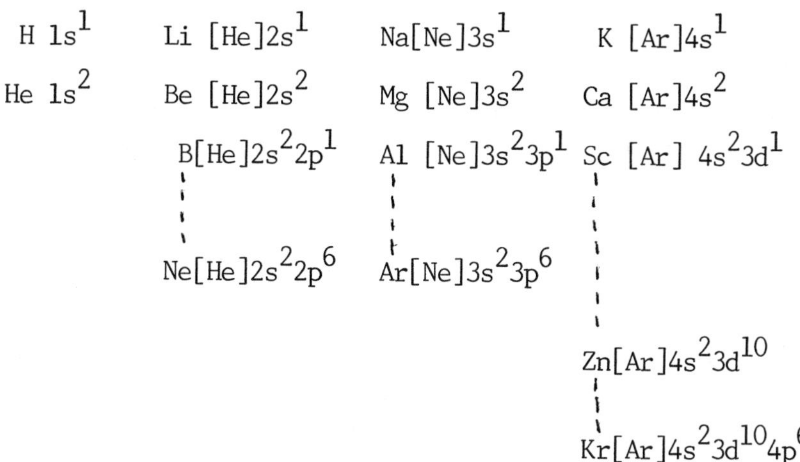

H $1s^1$ Li $[He]2s^1$ Na$[Ne]3s^1$ K $[Ar]4s^1$

He $1s^2$ Be $[He]2s^2$ Mg $[Ne]3s^2$ Ca $[Ar]4s^2$

B$[He]2s^2 2p^1$ Al $[Ne]3s^2 3p^1$ Sc $[Ar]$ $4s^2 3d^1$

Ne$[He]2s^2 2p^6$ Ar$[Ne]3s^2 3p^6$

Zn$[Ar]4s^2 3d^{10}$

Kr$[Ar]4s^2 3d^{10} 4p^6$

Beyond krypton, it's simplest to derive electron configurations from the periodic table.

Groups 1, 2: fill s sublevel
Groups 13-18: fill p sublevel
Groups 3-12: fill d sublevel (transition metals)
Lanthanides and actinides fill 4f, 5f levels

Electron configuration iodine atom? $[Kr]5s^2 4d^{10} 5p^5$

B. <u>3rd and 4th quantum numbers</u>

1. Orbital designated by m_ℓ = ℓ, - - - 0 - - -,-ℓ

ℓ = 0 (s sublevel); m_ℓ = 0 (one s orbital)

ℓ = 1 (p sublevel); m_ℓ = 1, 0, -1 (three p orbitals

ℓ = 2 (d sublevel); m_ℓ = 2, 1, 0, -1, -2 (5 d orbitals)

Each orbital has a capacity of two electrons. s orbitals are spherically symmetric about nucleus; p orbitals are dumbell shaped and are at right angles to each other.

2. Electrons in an orbital can have either of two spins; m_s = ½, -½

55

C. <u>Orbital diagram</u> Show number of electrons in each orbital and spin
of each electron.

	1s	2s	2p
H	(↑)		
He	(↑↓)		
Li	(↑↓)	(↑)	
Be	(↑↓)	(↑↓)	
B	(↑↓)	(↑↓)	(↑) ( ) ( )
C	(↑↓)	(↑↓)	(↑) (↑) ( )
N	(↑↓)	(↑↓)	(↑) (↑) (↑)
O	(↑↓)	(↑↓)	(↑↓) (↑) (↑)

Note that:

1. $2e^-$ in orbital have opposed spins

2. When several orbitals of same sublevel are available, e^- enter
   singly with parallel spins

Abbreviated electron configuration, orbital diagram of Fe?

$[Ar]4s^2 3d^6$

	4s	3d
$[Ar]$	(↑↓)	(↑↓) (↑) (↑) (↑) (↑)

LECTURE 3

I <u>Electronic Structure</u>

A. <u>Monatomic ions</u>

   1. Ions with noble gas structures (Groups 1, 2, 16, 17)

   2. Transition metal cations; outer s electrons are lost

$$_{24}Cr^{3+} \quad [Ar]3d^3 \qquad _{27}Co^{2+} \quad [Ar]3d^7 \qquad _{30}Zn^{2+} \quad [Ar]3d^{10}$$

II <u>Trends in Periodic Table</u>

A. <u>Atomic radius</u>
   1. In general, atomic radius decreases going across a period from
left to right, increases going down group.

     Na 0.186 nm   Mg 0.160 nm   S 0.104 nm   Cl 0.099 nm
     K 0.231 nm   Br 0.114 nm

These trends can be explained in terms of effective nuclear
charge felt by outer electron(s). Electrons in same outer

level do not shield one another effectively.

B. <u>Ionic radius</u>

Trends parallel those in atomic radius. Beyond that:

cations are smaller than the corresponding atoms
anions are larger than the corresponding atoms

This means that, in a typical ionic compound, the anions occupy most of the space.

C. <u>Ionization energy</u> = energy that must be absorbed to convert an atom to a +1 ion

$$Na(g) \longrightarrow Na^+(g) + e^- \qquad I.E. = +496 \text{ kJ/mol}$$

increases going across in period table, as atoms get smaller
decreases going down in periodic table, as atoms get larger

D. <u>Electronegativity</u> Measure of attraction of atom for electrons in covalent bond.

increases going across, decreases going down

DEMONSTRATIONS

1. Line spectra: GILB L 22; J. Chem. Educ. <u>69</u> 829 (1992)
2. Flame colors: GILB M 26, M 28; J. Chem. Educ. <u>67</u> 791 (1990), <u>68</u> 937 (1991), <u>69</u> 327 (1992), <u>71</u> 68 (1994), <u>72</u> 828 (1995)
3. Paramagnetism: GILB L 32
4. Oxidation states of vanadium: GILB M 264
5. Chemiluminescence: SHAK <u>1</u> 133

VIDEO

1. Wavelength and color: Falcon 6
2. Flame colors: Falcon 8
3. Hydrogen spectrum: Falcon 7
4. Paramagnetism: JCES 31
5. Chemiluminescence: SHAK 33

PROBLEMS

1. low frequency: low energy, high wavelength

 high frequency: high energy, low wavelength

3. a. $\lambda = \dfrac{2.998 \times 10^8 \text{ m/s}}{5.82 \times 10^{14}\text{/s}} = 5.15 \times 10^{-7}$ m $= 515$ nm

 b. $E = (6.626 \times 10^{-34} \text{ J·s})(5.82 \times 10^{14}\text{/s}) = 3.86 \times 10^{-19}$ J

 c. $E = 3.86 \times 10^{-19}$ J $\times 6.02 \times 10^{23} \times \dfrac{1 \text{ kJ}}{10^3 \text{ J}} = 232$ kJ

5. a. $\lambda = \dfrac{2.998 \times 10^8 \text{ m/s}}{2.001 \times 10^{13}\text{/s}} \times \dfrac{10^9 \text{ nm}}{1 \text{ m}} = 1.498 \times 10^4$ nm

 b. infrared

 c. $E = (6.626 \times 10^{-34} \text{ J·s})(2.001 \times 10^{13}\text{/s})(6.022 \times 10^{20} \text{ kJ/J·mol})$

 $= 7.984$ kJ/mol

7. a. $\Delta E = 675 \dfrac{\text{kJ}}{\text{mol}} \times \dfrac{1 \text{ mol}}{6.022 \times 10^{23}} \times \dfrac{10^3 \text{ J}}{1 \text{ kJ}} = 1.12 \times 10^{-18}$ J

 $\lambda = \dfrac{(6.626 \times 10^{-34} \text{ J·s})(2.998 \times 10^8 \text{ m/s})}{1.12 \times 10^{-18} \text{ J}} = 1.77 \times 10^{-7}$ m $= 177$ nm

 b. $\nu = \dfrac{2.998 \times 10^8 \text{ m/s}}{1.77 \times 10^{-7} \text{ m}} = 1.69 \times 10^{15}\text{/s}$

9. a. $\nu = \dfrac{2.998 \times 10^8 \text{ m/s}}{4.00 \times 10^{-9} \text{ m}} = 7.50 \times 10^{16}\text{/s}$

 b. $(6.626 \times 10^{-34} \text{ J·s})(7.50 \times 10^{16}\text{/s}) = 4.97 \times 10^{-17}$ J

11. a. (1)   b. (2)   c. (1)

13. a. transitions from $n_{hi} = 2, 3, 4, - -$ to $n_{lo} = 1$
 b. transitions from $n_{hi} = 3, 4, 5, - -$ to $n_{lo} = 2$

15. $\nu = \dfrac{2.180 \times 10^{-18}}{6.626 \times 10^{-34} \text{ s}}$ $(1/25 - 1/36) = 4.021 \times 10^{13}/s$

$\lambda = (2.998 \times 10^8 \text{ m/s})/(4.021 \times 10^{13}/s) = 7.456 \times 10^{-6} \text{ m} = 7456 \text{ nm}$

17. $\Delta E = \dfrac{(6.626 \times 10^{-34} \text{ J} \cdot s)(2.998 \times 10^8 \text{ m/s})}{97.23 \times 10^{-9} \text{ m}} = 2.043 \times 10^{-18} \text{ J}$

$2.043 \times 10^{-18} \text{ J} = 2.180 \times 10^{-18} \text{ J } (1 - x)$

$1 - x = 0.937; \quad x = 0.063; \quad n = 4$

19. Bohr model gives position of electron; quantum mechanical model gives only probabilities

21. a. 1, 0, -1   b. 2, 1, 0, -1, -2

c. $\ell = 2$; $m_\ell = 2, 1, 0, -1, -2$ $\ell = 1$; $m_\ell = 1, 0, -1$ $\ell = 0$; $m_\ell = 0$

23. a. 3p   b. 3d   c. 3d   d. 4f

25.   a. s   b. d   c. f

27. a. 4   b. 3   c, 5

29. a. not related except through ℓ

b. $m_\ell = +\ell, - - 0 - - -\ell$

c. none

31. a. cannot have a 1p sublevel   b. m_ℓ cannot be 2 when $\ell = 1$

e. m_s cannot be 0

33. a. $1s^2 2s^2 2p^6$ b. $1s^2 2s^2 2p^6 3s^2 3p^5$

c. $1s^2 2s^2 2p^6 3s^2 3p^6 4s^2 3d^7$

d. $1s^2 2s^2 2p^6 3s^2 3p^6 4s^2 3d^{10} 4p^6 5s^2 4d^{10}$

e. $1s^2 2s^2 2p^6 3s^2 3p^6 4s^2 3d^{10} 4p^6 5s^2 4d^{10} 5p^6 6s^1$

35. a. [Ar] $4s^1$ b. [Ar] $4s^2 3d^{10} 4p^5$ c. [Kr] $5s^2 4d^1$

d. [Xe] $6s^2 4f^{14} 5d^4$ e. [Xe] $6s^2 4f^{14} 5d^{10} 6p^2$

37. a. B   b. Nd   c. Zn   d. Mg

39. a. 7/13 = 0.538   b. 24/79 = 0.304   c. 18/47 = 0.383

41. a. excited   b. excited   c. ground
    d. impossible   e. excited   f. impossible

43.   1s   2s   2p   3s   3p   4s   3d

    a. (↑↓) (↑)

    b. (↑↓) (↑↓) (↑↓)(↑↓)(↑↓) (↑↓) (↑)(↑)(↑)

    c. (↑↓) (↑↓) (↑↓)(↑↓)(↑↓) (↑↓) (↑↓)(↑↓)(↑↓) (↑↓) (↑↓)(↑)(↑)(↑)(↑)

    d. (↑↓) (↑↓) (↑↓)(↑↓)(↑)

45. a. Mg   b. P   c. O

47. a. Kr and all elements of higher atomic number   b. Xe

    c. Si, Ge, As, Sb, Te, Po, At   d. Li, B, F

49. a. 3   b. 1   c. 6

51. a. Ca   b. K, Ga   c. none   d. none

53. a. $1s^2 2s^2 2p^6 3s^1$; $1s^2 2s^2 2p^6$

    b. $1s^2 2s^2 2p^6 3s^2 3p^4$; $1s^2 2s^2 2p^6 3s^2 3p^6$

    c. $1s^2 2s^2 2p^6 3s^2 3p^6 4s^2 3d^1$; $1s^2 2s^2 2p^6 3s^2 3p^6$

    d. $1s^2 2s^2 2p^6 3s^2 3p^6 3d^5$; $1s^2 2s^2 2p^6 3s^2 3p^6 3d^3$

55. a. 0   b. 2   c. 0   d. 0

57. a. Te < Sn < Sr   b. Sr < Sn < Te   c. Sr < Sn < Te

59. a. Sb   b. Cs   c. Cs

61. a. Ca   b. F^- c. Sr   d. Cu^+

63. a. K > Na > Mg > Mg^{2+} b. As^{3-} > As > P > N

65. $E = \dfrac{(6.626 \times 10^{-34} \text{ J·s})(2.998 \times 10^8 \text{ m})}{565 \times 10^{-9} \text{ m}} = 3.52 \times 10^{-19}$ J/particle

    $(25 \times 0.085 \text{ J/s})/(3.52 \times 10^{-19} \text{ J/particle}) = 6.0 \times 10^{18}$ particle/s

67. $E = \dfrac{(6.626 \times 10^{-34} \text{ J·s})(2.998 \times 10^8 \text{ m/s})}{10.6 \times 10^{-6} \text{ m}} = 1.87 \times 10^{-20}$ J/particle

    $(1 \text{ J/pulse})/(1.87 \times 10^{-20} \text{ J/particle}) = 5.35 \times 10^{19}$ particle/pulse

69. a. two electrons cannot have same four quantum numbers

60

b. maximum number of unpaired electrons

c. wavelength

d. n = 1, 2, 3, - -

71. a. true   b. - - n^2 rather than ℓ c. - - before rather than
" as soon as"

73. a. increased charge pulls electrons closer to nucleus

b. loses three e^- to give argon structure

c. elements become less metallic

75. a. chlorine, Cl   b. lead, Pb   c. manganese, Mn

d. lithium, Li   e. krypton, Kr

76.. Z = 3, n = 2

$$E = -2.180 \times 10^{-18} \text{ J} \times \frac{9}{4} \times \frac{1 \text{ kJ}}{10^3 \text{ J}} \times \frac{6.022 \times 10^{23}}{\text{mol}}$$

$$= -2.954 \times 10^3 \text{ kJ}$$

$$\Delta E = 0 - (-2954 \text{ kJ}) = 2954 \text{ kJ}$$

77. $\Delta E = 2.180 \times 10^{-18} \text{ J } (1/4 - 1/n^2)$

$$= (2.180 \times 10^{-18} \text{ J})(n^2 - 4)/4n^2$$

$$\lambda = \frac{(6.626 \times 10^{-34} \text{ J·s})(2.998 \times 10^8 \text{ m/s})}{2.180 \times 10^{-18}(n^2 - 4) \text{ J}} \times 4n^2 \times \frac{10^9 \text{ nm}}{1 \text{ m}}$$

$$= 364.5 \, n^2/(n^2 - 4)$$

78. n = 1; ℓ = 0; m_ℓ = 0, 1 $4e^-$
    ℓ = 1; m_ℓ = 0, 1, 2 $6e^-$

n = 2; ℓ = 0; m_ℓ = 0, 1 $4e^-$
    ℓ = 1; m_ℓ = 0, 1, 2 $6e^-$
    ℓ = 2; m_ℓ = 0, 1, 2, 3 $8e^-$

$1s^4 1p^4$

79. a. $3e^-$; $9e^-$; $15e^-$ b. $27e^-$ c. $1s^3 2s^3 2p^2$; $1s^3 2s^3 2p^9 3s^2$

80. a. 540 nm; $E = \dfrac{(6.626 \times 10^{-34} \text{ J} \cdot \text{s})(2.998 \times 10^{8} \text{m/s})}{540 \times 10^{-9} \text{ m}}$

$$= 3.68 \times 10^{-19} \text{ J}$$

400 nm: $E = 4.97 \times 10^{-19} \text{ J}$

$E_{min} = 3.68 \times 10^{-19} \text{ J} - 0.26 \times 10^{-19} \text{ J} = 3.42 \times 10^{-19} \text{ J}$

b. with no kinetic energy:

$$= \dfrac{(6.626 \times 10^{-34} \text{ J} \cdot \text{s})(2.998 \times 10^{8} \text{ m/s})}{3.42 \times 10^{-19} \text{ J}}$$

$$= 5.81 \times 10^{-7} \text{ m} = 581 \text{ nm}$$

CHAPTER 7
Covalent Bonding

LECTURE NOTES

This chapter requires three lectures. The first lecture is devoted primarily to Lewis structures. The second lecture covers molecular geometry and polarity. The final lecture deals with hybridization, sigma and pi bonding. Note that

1. Students find Lewis structures relatively simple to draw provided they get sufficient practice; a molecular model kit helps. Note that the ability to write Lewis structures is essential to an understanding of just about everything else in the chapter, including molecular geometry and hybridization.

2. You may or may not want to discuss formal charge (p. 180). It is useful in choosing between alternative skeletons, but is also somewhat of a diversion in a closely-packed chapter.

3. Table 7.3 is a useful way of summarizing the geometries of all species in which a central atom is surrounded by 2, 3 or 4 electron pairs. The AXE notation is useful because it enables students to predict the geometries of multiple-bonded species as well.

4. Figure 7.7 shows the geometries of species where the central atom is surrounded by 5 or 6 electron pairs, including unshared pairs. In principle, these geometries can all be predicted by VSEPR theory; in practice a certain amount of memorization is necessary.

5. Students often get the mistaken idea that promotion of electrons is necessary for hybridization to occur. They also have trouble distinguishing between sigma and pi bonds.

<u>LECTURE 1</u>

I <u>Lewis Structures</u>

A Lewis structure shows the distribution of outer (valence) electrons in an atom, molecule or polyatomic ion. Unshared electrons are shown as dots, bonds as straight lines.

$$H\cdot \quad + \quad \cdot \ddot{\underset{\cdot\cdot}{F}}{:} \quad \longrightarrow \quad H - \ddot{\underset{\cdot\cdot}{F}}{:}$$

$$2H\cdot \quad + \quad \cdot \ddot{\underset{\cdot\cdot}{O}}\cdot \quad \longrightarrow \quad H - \underset{\cdot\cdot}{\ddot{O}} - H$$

In H_2O and HF, as in most molecules and polyatomic ions, nonmetal atoms except H are surrounded by 8 electrons, an octet. In this sense, each atom has a noble gas structure. Lewis structures are written following a stepwise procedure.

A. <u>Rules for writing Lewis structures</u> (single bonds)

    1. Count valence electrons available. Number of valence electrons contributed by nonmetal atom is equal to the last digit of its group number in the periodic table (1 for H). Add electrons to take into account negative charge.

        OCl^- ion: $6 + 7 + 1 = 14$ valence e^-

        CH_3OH molecule: $4 + 4(1) + 6 = 14$ valence e^-

        SO_3^{2-} ion: $6 + 3(6) + 2 = 26$ valence e^-

    2. Draw skeleton structure, using single bonds

$$O - Cl \qquad H - \overset{\displaystyle H}{\underset{\displaystyle H}{\overset{|}{\underset{|}{C}}}} - O - H \qquad O - \overset{}{\underset{\displaystyle O}{\overset{}{\underset{|}{S}}}} - O$$

        Note that carbon almost always forms four bonds. Central atom is written first in formula. Terminal atoms are most often H, O, or a halogen.

    3. Deduct two electrons for each single bond in the skeleton.

        OCl^- ion: $14 - 2 = 12$ valence e^- left

        CH_3OH molecule: $14 - 10 = 4$ valence e^- left

        SO_3^{2-} ion: $26 - 6 = 20$ valence e^- left

    4. Distribute these electrons to give each atom a noble gas structure, if possible.

$$(\,{:}\ddot{\underset{\cdot\cdot}{O}} - \ddot{\underset{\cdot\cdot}{C}}l{:}\,)^- \qquad H - \overset{\displaystyle H}{\underset{\displaystyle H}{\overset{|}{\underset{|}{C}}}} - \ddot{\underset{\cdot\cdot}{O}} - H \qquad (\,{:}\ddot{\underset{\cdot\cdot}{O}} - \overset{}{\underset{\displaystyle {:}\ddot{\underset{\cdot\cdot}{O}}{:}}{\overset{}{\underset{|}{S}}}} - \ddot{\underset{\cdot\cdot}{O}}{:}\,)^{2-}$$

64

B. <u>Too few electrons</u>; form multiple bonds   Structure of NO_3^- ion?

number of valence e^- = 5 + 18 + 1 = 24

skeleton   O – N – O
                     |
                     O

valence e^- left = 24 – 6 = 18.

If all electrons are distributed as unshared pairs

$$:\!\ddot{O} - N - \ddot{O}:$$
$$|$$
$$:\ddot{O}:$$

Nitrogen atom is surrounded by only 6 valence e^-. To remedy a deficiency of two electrons, form a double bond.

$$:\!\ddot{O} - N - \ddot{O}:$$
$$\|$$
$$:O:$$

Structure of N_2?   10 valence e^- $:N \equiv N:$

C. <u>Too many electrons</u>   Give central atom an expanded octet

Consider XeF_4, 36 valence e^-. Octet structure uses 32 e^-

In a few molecules, there are less than 8 electrons around the central atom.

$$:\ddot{F} - Be - \ddot{F}: \qquad :\ddot{F} - B - \ddot{F}:$$
$$|$$
$$:F:$$

D. <u>Resonance forms</u>   To explain the fact that all three bonds in the nitrate ion are identical, invoke the concept of resonance.

$$:\ddot{O} = N - \ddot{O}: \longleftrightarrow :\ddot{O} - N - \ddot{O}: \longleftrightarrow :\ddot{O} - N = \ddot{O}:$$
$$|\qquad\qquad\qquad \|\qquad\qquad\qquad |$$
$$:\ddot{O}:\qquad\qquad\quad :O:\qquad\qquad\quad :\ddot{O}:$$

True structure is a hybrid of these three forms. Note that

1. Resonance forms obtained by moving electrons, not atoms
2. Resonance can be expected when it is possible to draw more than
   one structure that follows the octet rule.

<u>LECTURE 2</u>

I <u>Molecular Geometry</u>

VSEPR principle: Electron pairs around a central atom tend to be oriented so as to be as far apart as possible.

A. <u>Two to six atoms around central atom;</u> no unshared pairs

   BeF_2 linear, BF_3 triangular planar, CF_4 tetrahedral, PF_5 triangular
   bipyramid, SF_6 octahedral. Discuss bond angle.

B. <u>Unshared pairs</u> (Table 7.3)

   AX_2E (GeF_2): bent, $120°$
   AX_3E (NH_3): triangular pyramid, $109°$
   AX_2E_2 (H_2O): bent, $109°$
   Expanded octets: refer to Figure 7.7

C. <u>Multiple bonding</u> Has no effect upon geometry; Table 7.3 applies

   BF_3 and SO_3: both AX_3 molecules, same geometry
   BeF_2 and CO_2: both AX_2, both linear

II <u>Polarity</u>

A. <u>Bond polarity</u> All bonds are polar unless the two atoms joined
   <u>are identical</u> (H - H). Extent of polarity depends upon difference
   in electronegativity.

   H - H Δ E.N. = 0   nonpolar

   H - C Δ E.N. = 0.3   slightly polar

   H - F Δ E. N. = 1.8   strongly polar (- pole at F atom)

B. <u>Molecular polarity</u>

   1. Diatomic molecules: polar if atoms differ

      H - Cl   Cl - Cl
      polar   nonpolar

   HCl molecules line up in electric field; Cl_2 molecules don't

   2. Polyatomic molecules. Even though bonds are polar, molecule
      may be nonpolar if bonds are symmetrically arranged.

66

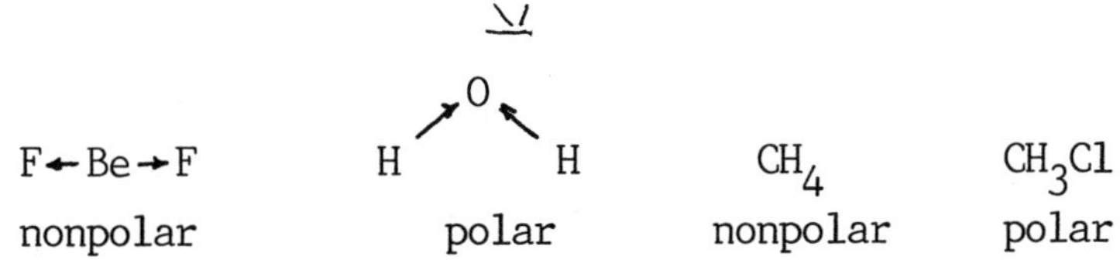

LECTURE 3

I Atomic Orbitals; Hybridization

In molecules, the orbitals occupied by electron pairs are seldom "pure" s or p orbitals. Instead, they are "hybrid" orbitals, formed by combining s, p and d orbitals.

A. Formation of hybrid orbitals

1. s orbital + p orbital $\rightarrow$ two sp hybrid orbitals

Be in BeF_2 2s (↿⇂) 2p (↿⇂)

2. s orbital + two p orbitals $\rightarrow$ three sp^2 hybrid orbitals

B in BF_3 2s (↿⇂) 2p (↿⇂)(↿⇂)

3. s orbital + three p orbitals $\rightarrow$ four sp^3 hybrid orbitals

C in CH_4 2s (↿⇂) 2p (↿⇂)(↿⇂)(↿⇂)

B. Hybridization with 5 or 6 electron pairs: sp^3d, sp^3d^2. Note that expanded octets do not occur with atoms in the second period (e.g., N, O, F) since there are no 2d orbitals.

C. Unshared pairs can be hybridized as in ammonia and water. Only one of the electron pairs in a multiple bond is hybridized.

$$:\ddot{O} - \ddot{S} = \ddot{O}:\qquad sp^2 \text{ hybridization for sulfur}$$

$$:\ddot{O} = C = \ddot{O}:\qquad sp \text{ hybridization for carbon}$$

II Sigma and Pi Bonds

When a bond consists of an electron pair in a hybrid orbital, the electron density is concentrated along the bond axis and is symmetrical about it. Such a bond is called a sigma bond.

The "extra" electron pairs in a multiple bond are located in unhybridized orbitals which are not concentrated along the bond axis. Such bonds are called pi bonds.

CO_2 two sigma bonds, two pi bonds

SO_2 two sigma bonds, one pi bond

N_2 one sigma bond, two pi bonds

DEMONSTRATIONS

1. Nitrogen oxide and nitrogen dioxide: GILB D 27; SHAK 1 117, 2 163; J. Chem. Educ. 68 938 (1991)
2. Paramagnetism of oxygen: GILB M 190; SHAK 2 147; J. Chem. Educ. 57 373 (1980)
3. Molecular geometry with balloons: GILB E 18
4. Reaction of xenon with fluorine: J. Chem. Educ. 43 202 (1966)

VIDEO

1. Reaction of NO with carbon disulfide: SHAK 7
2. Paramagnetism of oxygen: SHAK 38; JCES 30

PROBLEMS

1. a. Br–C(–Br)(–Br)–Br   b. Cl–N(–Cl)–Cl   c. $(:N \equiv O:)^+$

    d. $(:\ddot{O} - \ddot{C}l - \ddot{O}:)^-$

3. a. $(:\ddot{C}l - P(–Cl)(–Cl) - \ddot{C}l:)^+$ b. $(:\ddot{F} - \ddot{C}l - \ddot{F}:)^+$ c. $(:\ddot{I} - \ddot{I} - \ddot{I}:)^-$

    d. F–Si(–F)(–F)–F

5. a. $:C \equiv O:$ b. H–N(–H)–H

5. c. 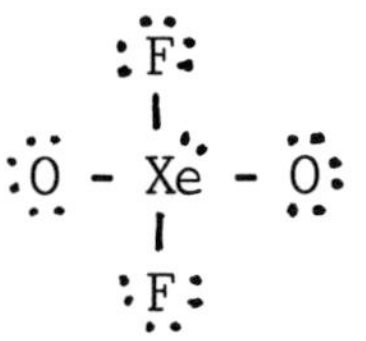 Lewis structure: O–Xe–O with F above and F below Xe; each O has dots, each F has dots (:F:)

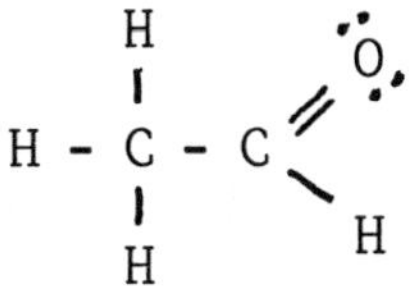

    d. 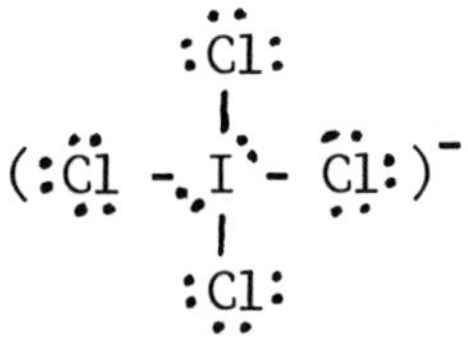 (:Cl – I – Cl:)⁻ with :Cl: above and :Cl: below I; each Cl with dots

7. 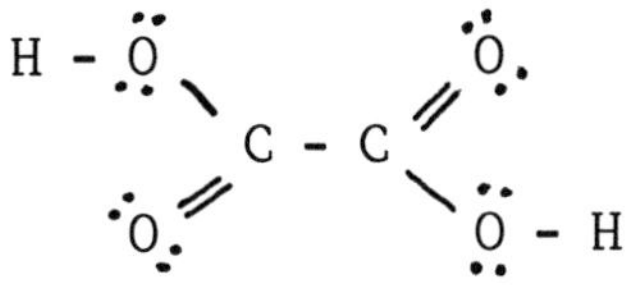 H – O
        \
         C – C
        /   \\
     O   O – H
    (oxalic acid Lewis structure with C–C, C=O double bonds, O–H groups)

9. a. 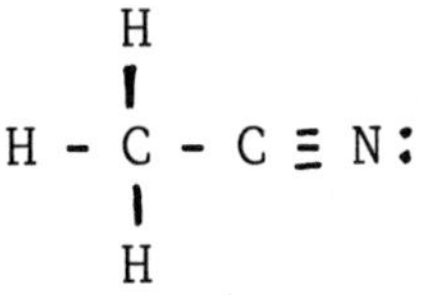
```
        H
        |
   H -  C - C ≡ N:
        |
        H
```

   b.
```
        H       O
        |      ‖
   H -  C -  C
        |      \
        H       H
```

   c.
```
      :F:  :Cl:
       |    |
  :F - C =  C - Cl:
```

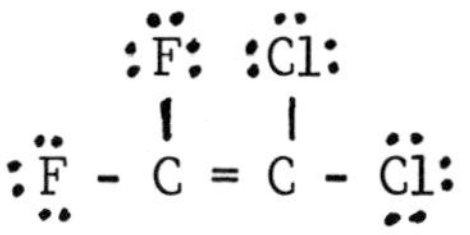

11. 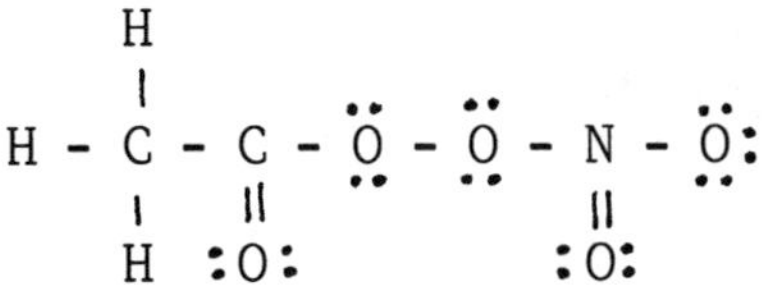
```
        H
        |
   H -  C - C - O - O - N - O:
        |   ‖           ‖
        H  :O:         :O:
```

13. 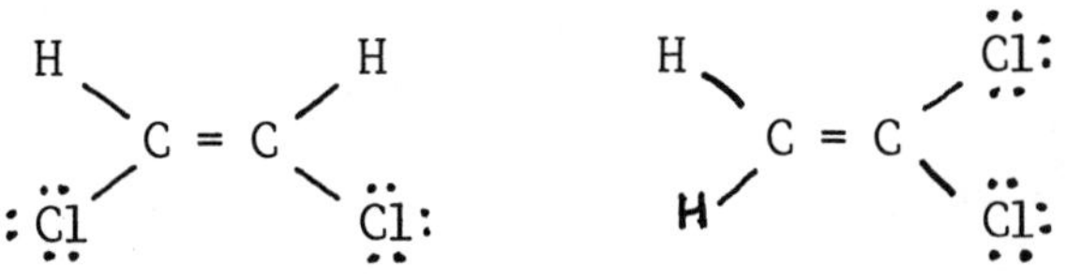
```
    H          H              H         Cl:
     \        /                \       /
      C  =  C                   C  =  C
     /        \                /       \
  :Cl         Cl:            H          Cl:
```

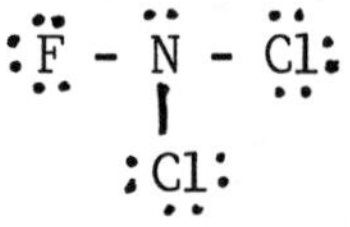

15. a. HCl   b. Cl_2 c. CO   d. CCl_4

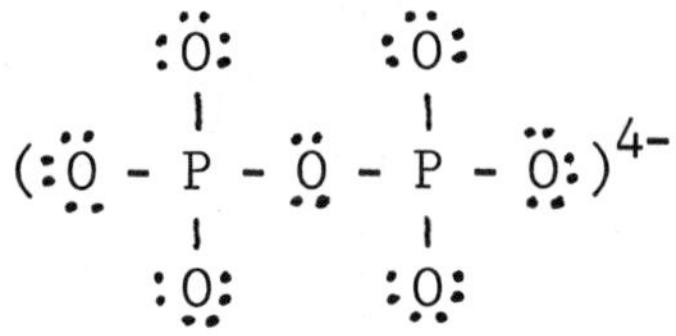

17. a. 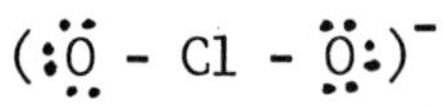 (:O – Cl – O:)⁻   b. 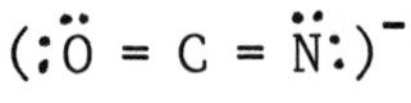 (:O = C = N:)⁻   c.
```
         :F - N - Cl:
              |
             :Cl:
```

```
            :O:       :O:
             |         |
   d. ( :O - P - O - P - O: )⁴⁻
             |         |
            :O:       :O:
```

19. a. H – Be – H   b. 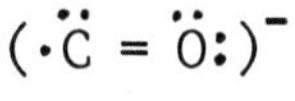 (·C 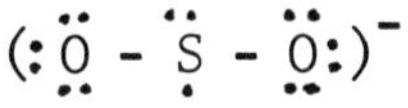 = O:)⁻   c. (:O – S – O:)⁻   d. 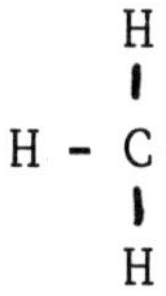
```
        H
        |
   H -  C·
        |
        H
```

21. a. 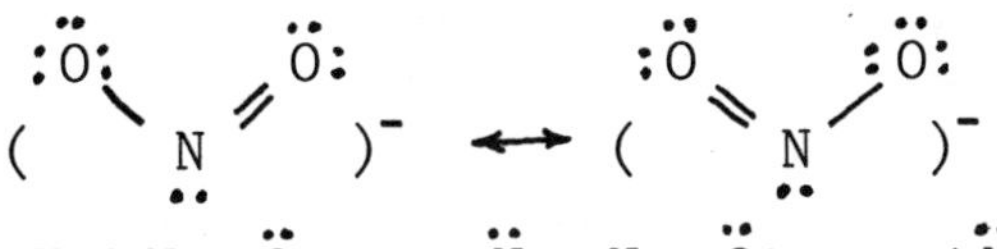
```
     :O       O:                :O       :O:
       \     ‖                     ‖     /
        N             ↔             N
                 )⁻                       )⁻
   (          )               (          )
```

   b. 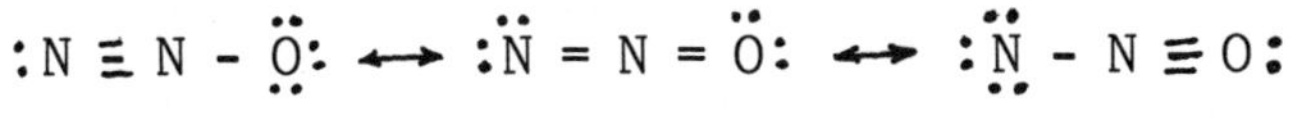 :N ≡ N – O: ↔ :N = N = O: ↔ :N – N ≡ O:

   c. 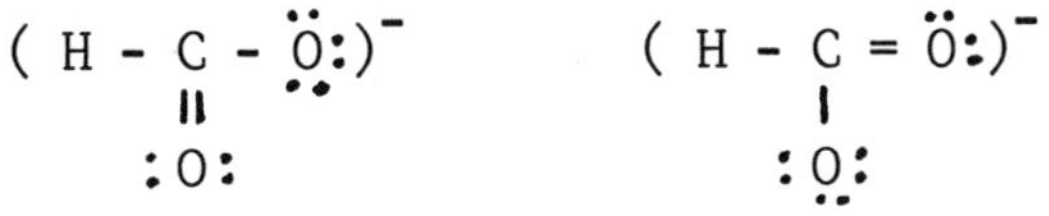 (H – C – O:)⁻   (H – C = O:)⁻
```
           ‖                            |
          :O:                          :O:
```

69

23. a. H – N ≡ N – N̈: , H – N̈ – N ≡ N: b. no; isomer

25.

$$
\begin{array}{ccc}
& \overset{\textstyle H}{\underset{\textstyle |}{B}} & \\
H-N & & N-H \\
\| & | & \\
H-B & & B-H \\
& \underset{\textstyle |}{\overset{\textstyle N}{}} & \\
& H &
\end{array}
\longleftrightarrow
\begin{array}{ccc}
& \overset{\textstyle H}{\underset{\textstyle |}{B}} & \\
H-N & & N-H \\
| & & \| \\
H-B & & B-H \\
& \underset{\textstyle |}{\overset{\textstyle N}{}} & \\
& H &
\end{array}
$$

27. a. $(:\ddot{N} = N = \ddot{N}:)^-$; +1   b. [XeF₆ structure] ; 0

c. :Cl̈ – B̈r – Cl̈: ; 0
              |
            :Cl̈:

29. a. H – C ≡ N: H = 0, C = 0, N = 0   more likely

    H – N ≡ C: H = 0, C = −1, N = +1

 b. :N̈ = Ö – Cl̈:   N = −1, O = +1, Cl = 0

    :Ö = N̈ – Cl̈:   O = 0, N = 0, Cl = 0   more likely

31. a. [SO₂ structure]   b. :Cl̈ – Be – Cl̈:   c. :Cl̈ – S̈e – Cl̈:
          bent   linear |
                                                            :Cl̈:

    d. triangular bipyramid   see-saw

33. a. :N̈ = N = Ö: b. :Ö = N̈ – Cl̈:   c. tetrahedral

          linear   bent

    d. $(:\ddot{O} – \ddot{O} – \ddot{O}:)^{2-}$

          bent

35. a. tetrahedral   b. see-saw (:Br - Te - Br: with :Br: above and :Br: below)   c. T-shaped (:F - I - F: with :F: below)

d. octahedral

37. a square planar   b. triangular bipyramid   c. bent

39. a. 180°   b. 120°   c. 109.5°, 120°   d. 109.5°, 120°

41. a.

$$H - \overset{\overset{\displaystyle H}{|}}{\underset{\underset{\displaystyle H}{|}}{C}} - \overset{}{\underset{\underset{\displaystyle :O:}{||}}{C}} - \ddot{O} - \ddot{O} - \overset{}{\underset{\underset{\displaystyle :O:}{||}}{C}} - \overset{\overset{\displaystyle H}{|}}{\underset{\underset{\displaystyle H}{|}}{C}} - H$$

b. 109.5° around CH_3 carbons, 120° around C = O, 109.5° around central O

43. PH_3 and H_2S; unshared pairs on P, S

45. 1: 120°   2: 109.5°   3: 109.5°

47. a

49. a. polar   b. polar   c. nonpolar   d. nonpolar

51. 1st molecule is polar; dipoles do not cancel

53. a. sp^2 b. sp   c. sp^3d d. sp^3d^2

55. a. sp   b. sp^2 c. sp^3 d. sp^3

57. a. sp^3 b. sp^3d c. sp^3d d. sp^3d^2

59. a. $(:F - Xe - F:)^-$ sp^3d^2 b. sp^3d^2 c. $(:Cl - P - Cl:)^-$ with :Cl: above and Cl below, sp^3d

61. sp^2

63. a. sp^3 b. sp^2 c. sp   d. sp^2

65. a. $H_2N - \overset{}{\underset{\underset{\displaystyle O}{||}}{C}} - NH_2$ sp^2 b. $F - \overset{}{\underset{\underset{\displaystyle O}{||}}{C}} - F$ sp^2 c. sp^3

67. a. 4 sigma   b. 3 sigma, 1 pi   c. 2 sigma, 2 pi   d. 3 sigma, 1 pi

69. 12 sigma, 3 pi

71. a. NH_3 b. N_2 c. NO_2^-

73. $:\ddot{Cl} = Be = \ddot{Cl}:$ $Cl = +1$, $Be = -2$

other structure has zero formal charges

75. 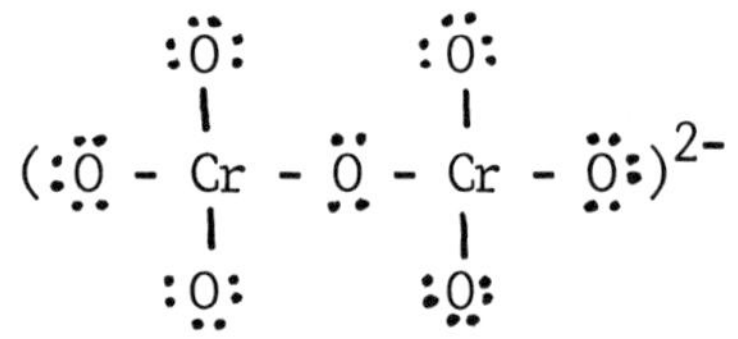 $$(:\ddot{O} - \underset{\overset{\displaystyle :O:}{|}}{\overset{\overset{\displaystyle :\ddot{O}:}{|}}{Cr}} - \ddot{O} - \underset{\overset{\displaystyle :O:}{|}}{\overset{\overset{\displaystyle :\ddot{O}:}{|}}{Cr}} - \ddot{O}:)^{2-}$$

77. b, d

79. AX_2E_2; 2; 2; bent; sp^3; polar

AX_3; 3; 0; triangular planar; sp^2; nonpolar

AX_4E_2; 4; 2; square planar; sp^3d^2; nonpolar

AX_5; 5; 0; triangular bipyramid; sp^3d; nonpolar

80. n chlorine fluoride $= \dfrac{(3.00 \text{ atm})(0.457 \text{ L})}{(0.0821 \text{ L·atm/mol·K})(348 \text{ K})} = 0.0480$ mol

n $UF_6 = 5.63$ g $\times \dfrac{1 \text{ mol}}{352.0 \text{ g}} = 0.0160$ mol

3 mol chlorine fluoride, 1 mol UF_6

$3ClF_3(g) + U(s) \longrightarrow UF_6(s) + 3ClF(g)$ $x = 3$

Lewis structure: $:\ddot{F} - \underset{\overset{}{..}}{\overset{\overset{\displaystyle :\ddot{F}:}{|}}{Cl}} - \ddot{F}:$

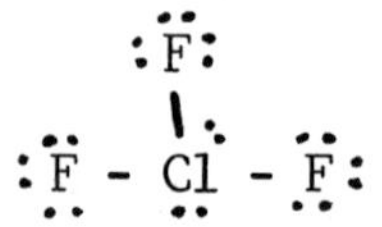

T-shaped; polar; 90°, 180°; sp^3d; 3 sigma

81. 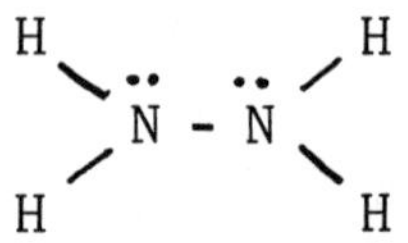 bent, 109.5° bond angle, polar

82. 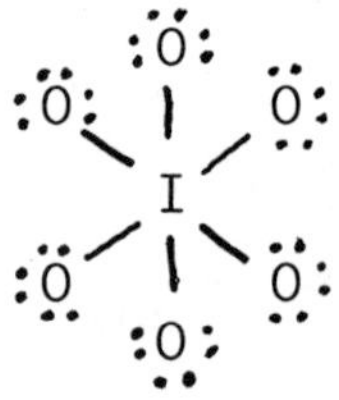 6, sp^3d^2, octahedral

83. a.

$$\begin{pmatrix} & :\overset{..}{O}: & \\ & | & \\ :\overset{..}{\underset{..}{O}} - & S - & \overset{..}{\underset{..}{O}}: \\ & | & \\ & :\overset{..}{\underset{..}{O}}: & \end{pmatrix}^{2-} \qquad \begin{pmatrix} & :O: & \\ & \| & \\ :\overset{..}{\underset{..}{O}} - & S - & \overset{..}{\underset{..}{O}}: \\ & \| & \\ & :O: & \end{pmatrix}^{2-}$$

b. tetrahedral   c. sp^3

d. 1st structure: $S = +2$, $O = -1$

    2nd structure: $S = 0$, $O = -1, -1, 0, 0$

84. a.

$$:\overset{\textstyle :\overset{..}{O}:}{\underset{\textstyle :\overset{..}{\underset{..}{Cl}}:}{\overset{|}{\underset{|}{Cl} - P - \overset{..}{\underset{..}{Cl}}:}}} \qquad Cl = 0,\ O = -1,\ P = +1$$

b.

$$:\overset{\textstyle :O:}{\underset{\textstyle :\overset{..}{\underset{..}{Cl}}:}{\overset{\|}{\underset{|}{Cl} - P - \overset{..}{\underset{..}{Cl}}:}}}$$

CHAPTER 8
Thermochemistry

LECTURE NOTES

Typically, students find this chapter difficult. In part, this reflects the fact that this is new material, seldom covered in high school. Moreover, thermochemistry is an abstract subject; students have trouble relating it to the real world. For this reason, the experimental aspects of thermochemistry, i.e., calorimetry, are presented early in the chapter.

In discussing calorimetry (Section 8.2), it is important to point out that

$$q_{reaction} = -q_{calorimeter}$$

i.e., that these two heat flows are equal in magnitude but opposite in sign. Students find the idea of the "calorimeter constant" C hard to grasp. It may help to relate C to the masses and specific heats of the bomb and the water it contains.

Thermochemical equations (Section 8.4) extend the conversion factor approach to heat flow. It is important to emphasize that ΔH is directly proportional to amount of reactant or product. Hess' law is not heavily stressed in this chapter; instead we emphasize the relation

$$\Delta H^\circ = \sum \Delta H^\circ_f \text{ products } - \sum \Delta H^\circ_f \text{ reactants}$$

We have shortened the discussion of bond energy. In particular, we no longer try to calculate ΔH from bond energies. In the real world, it is always obtained from heat of formation data.

Many instructors delete the discussion of the First Law (Section 8.7); it is not required as background in any subsequent chapter. If you cover it, keep in mind that college freshmen seldom appreciate the elegant logic of thermodynamics. Try to relate ΔE and ΔH to real processes.

This chapter requires a minimum of two lectures and could easily expand to three if you spend much time on Section 8.7. In the outline that follows, we strike an average and include material for $2\frac{1}{2}$ lectures.

<u>LECTURE 1</u>

I <u>Thermochemistry</u>

A. <u>Basic concepts</u>
Define and illustrate system, surroundings, state property. Basic
equation for heat flow:

$$q = c \times m \times \Delta t \qquad (c = \text{specific heat})$$

Suppose 652 J of heat is added to 15.0 g of water ($c = 4.18$ J/g°C),
originally at 20.0°C. Final t?

$$\Delta t = \frac{652 \text{ J}}{4.18 \text{ J/g} \cdot \text{°C} \times 15.0 \text{ g}} = 10.4\text{°C}; \text{ final } t = 30.4\text{°C}$$

Note that if heat is absorbed by system (q positive), temperature
increases; if q is negative, temperature drops. For a reaction
taking place at constant P and T:

endothermic: $q = \Delta H > 0$; reaction system absorbs heat
exothermic: $q = \Delta H < 0$; reaction system evolves heat

B. <u>Calorimetry</u>

1. Coffee-cup calorimeter ΔH reaction $= -q$ water. Heat given off
by reaction is absorbed by water in coffee cup.

Suppose heat is absorbed by 412 g of water, increasing its temperature from 20.12 to 29.86°C. What is ΔH?

$$q \text{ water} = 4.18 \frac{\text{J}}{\text{g} \cdot \text{°C}} \times 412 \text{ g} \times 9.74\text{°C} = 1.68 \times 10^3 \text{ J}$$

$$\Delta H = -1.68 \text{ kJ}$$

2. Bomb calorimter

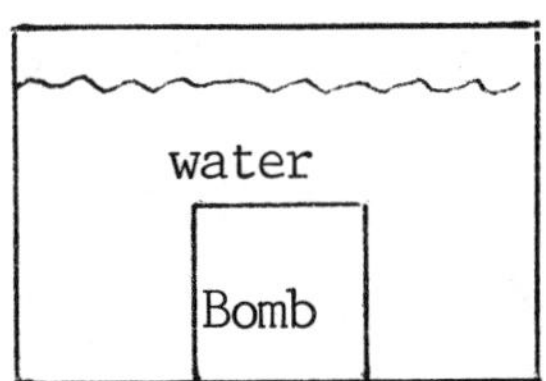

some heat is absorbed by
metal as well as water

$$q \text{ reaction} = -q \text{ calorimeter} = -(C \text{ calorimeter}) \times \Delta t$$

where C calorimter is the total heat capacity of bomb + water

Suppose combustion of 1.60 g of methane in bomb calorimeter raises temperature by 5.14°C (C = 17.2 kJ/°C)

$$q \text{ reaction} = -17.2 \text{ kJ/°C} \times 5.14°C = -88.4 \text{ kJ}$$

C. <u>Thermochemical Equations</u>   Specify ΔH in kilojoules

$$H_2(g) + Cl_2(g) \longrightarrow 2HCl(g) \qquad \Delta H = -185 \text{ kJ}$$

185 kJ of heat evolved when two moles of HCl are formed

$$2HgO(s) \longrightarrow 2Hg(l) + O_2(g) \qquad \Delta H = +182 \text{ kJ}$$

182 kJ of heat must be absorbed to decompose 2 mol HgO

LECTURE 2

I <u>Thermochemistry</u>

A. <u>Rules of thermochemistry</u>

1. ΔH is directly proportional to amount of reactants or products. When one mole of ice melts, 6.00 kJ of heat is absorbed, ΔH = +6.00 kJ. If one gram of ice melts, ΔH = 6.00 kJ/18.02 = +0.333 kJ. In general, ΔH can be related to amount by the conversion factor approach.

$$H_2(g) + Cl_2(g) \longrightarrow 2HCl(g) \qquad \Delta H = -185 \text{ kJ}$$

when 1.00 g of Cl_2 reacts:

$$\Delta H = 1.00 \text{ g } Cl_2 \times \frac{1 \text{ mol } Cl_2}{70.90 \text{ g } Cl_2} \times \frac{-185 \text{ kJ}}{1 \text{ mol } Cl_2} = -2.61 \text{ kJ}$$

2. ΔH for a reaction is equal in magnitude but opposite in sign to ΔH for the reverse reaction.

$$H_2O(s) \longrightarrow H_2O(l) \qquad \Delta H = +6.00 \text{ kJ; } 6.00 \text{ kJ absorbed}$$

$$H_2O(l) \longrightarrow H_2O(s) \qquad \Delta H = -6.00 \text{ kJ; } 6.00 \text{ kJ evolved}$$

3. Hess' law: If Equation 1 + Equation 2 = Equation 3, then

$$\Delta H_3 = \Delta H_1 + \Delta H_2$$

Often used to calculate ΔH for one step, knowing ΔH for all other steps and for the overall reaction.

76

$$C(s) + \tfrac{1}{2}O_2(g) \longrightarrow CO(g) \qquad \Delta H_1 = ?$$

$$CO(g) + \tfrac{1}{2}O_2(g) \longrightarrow CO_2(g) \qquad \Delta H_2 = -283.0 \text{ kJ}$$

$$C(s) + O_2(g) \longrightarrow CO_2(g) \qquad \Delta H_3 = -393.5 \text{ kJ}$$

$$\Delta H_1 = -110.5 \text{ kJ}$$

B. Heats of formation

1. Meaning. ΔH_f° of compound = ΔH when one mole of compound is formed from the elements in their stable states.

$$2Ag(s) + Cl_2(g) \longrightarrow 2AgCl(s) \qquad \Delta H^\circ = -254.0 \text{ kJ}$$

$$\Delta H_f^\circ \ AgCl(s) = -127.0 \text{ kJ}$$

$$HgO(s) \longrightarrow Hg(l) + \tfrac{1}{2}O_2(g) \qquad \Delta H^\circ = +90.8 \text{ kJ}$$

$$\Delta H_f^\circ \ HgO(s) = -90.8 \text{ kJ}$$

2. Usefulness. For any thermochemical equation:

$$\Delta H^\circ = \sum \Delta H_f^\circ \text{ products} - \sum \Delta H_f^\circ \text{ reactants}$$

Take heat of formation of element in stable state to be 0

$$CH_4(g) + 2\,O_2(g) \longrightarrow CO_2(g) + 2H_2O(l)$$

$$\Delta H^\circ = \Delta H_f^\circ \ CO_2(g) + 2\,\Delta H_f^\circ \ H_2O(l) - \Delta H_f^\circ \ CH_4(g)$$

$$= -393.5 \text{ kJ} + 2(-285.8 \text{ kJ}) - (-74.8 \text{ kJ}) = -890.3 \text{ kJ}$$

3. Can apply to ions, setting $\Delta H_f^\circ \ H^+(aq) = 0$

$$Zn(s) + 2H^+(aq) \longrightarrow Zn^{2+}(aq) + H_2(g)$$

$$\Delta H^\circ = \Delta H_f^\circ \ Zn^{2+}(aq) = -152.4 \text{ kJ}$$

LECTURE 2½

I Bond Energies

B.E. = ΔH when one mole of bonds is broken in gas state

$$Cl_2(g) \longrightarrow 2Cl(g) \qquad \Delta H = \text{B.E. Cl-Cl} = 243 \text{ kJ}$$

$$N_2(g) \longrightarrow 2N(g) \qquad \Delta H = \text{B.E. } N \equiv N = 941 \text{ kJ}$$

In general, multiple bonds are stronger than single bonds.

$$C-C = 347 \text{ kJ} \qquad C=C = 612 \text{ kJ} \qquad C-C = 820 \text{ kJ}$$

Saunders College Publishing

II $\underline{\text{First Law; } \Delta H \text{ and } \Delta E}$

    A. $\Delta E = q + w$ where ΔE = change in energy of system, q = heat flow into system, w = work done on system

    B. $\underline{\text{Constant volume, constant pressure processes}}$

        constant volume: $w = 0$; $q_V = \Delta E$

        constant pressure: $w = -P\Delta V$

$$q_p = \Delta H = \Delta E + \Delta n_g RT$$

DEMONSTRATIONS

1. Specific heats of liquids: GILB H 27

2. Exothermic reactions: GILB J 9

3. Endothermic solution process: GILB H 18, H 23; SHAK $\underline{1}$ 10

4. Chemical hot pack: SHAK $\underline{1}$ 36

5. Cold packs: SHAK $\underline{1}$ 8

6. Heats of solution: J. Chem. Educ. $\underline{67}$ 426 (1990)

7. Thermite reaction: GILB H 9, H 11; SHAK $\underline{1}$ 85

8. Reaction of hydrogen with oxygen: GILB A 20, A 21, M 14; SHAK $\underline{1}$ 106; J. Chem. Educ. 72, 177, 1128 (1995)

9. Reaction of hydrogen with chlorine: SHAK $\underline{1}$ 121

VIDEO

1. Endothermic reaction: JCES 16

2. Thermite reaction: SHAK 20; JCES 15; Falcon 32

3. Reaction of hydrogen with oxygen: SHAK 21; JCES 11

PROBLEMS

1. $q = 5.88 \text{ g} \times 5.34°C \times 0.523 \text{ J/g·}°C = 16.4 \text{ J}$

3. $375 = 1.8t + 32$; $t = 343/1.8 = 191°C$

    $q = 473 \text{ g} \times 168°C \times 0.702 \text{ J/g·}°C = 7.17 \times 10^4 \text{ J}$

5. a. $CaCl_2(s) \longrightarrow Ca^{2+}(aq) + 2Cl^-(aq)$

b. $q = -q$ water $= -150.0$ g x $1.75°C$ x 4.18 J/g·°C $= -1.10$ x 10^3 J

c. exothermic

d. $\mathcal{M} = 111.0$ g/mol; q water $= 1.10$ x 10^3 J x $\dfrac{111.0}{1.50} = 8.14$ x 10^4 J

7. q for sample $= -2.115$ x 10^4 J/°C x $4.22°C$

$\mathcal{M} = 180.2$ g/mol

q per mole $= -2.115$ x 10^4 J/°C x $4.22°C$ x $\dfrac{180.2}{4.50} = -3.57$ x 10^6 J

9. $q = \dfrac{4.96 \text{ x } 10^3 \text{ kJ}}{194.2 \text{ g}}$ x 2.500 g $= 63.9$ kJ

$C = 63.9$ kJ/$4.59°C = 13.9$ kJ/°C

11. $q = \dfrac{726 \text{ kJ x } 5(0.796)}{32.04} = 90.2$ kJ

$\Delta t = 90.2$ kJ/$(8.342$ kJ/°C$) = 10.8°C$; $t_{final} = 33.5°C$

13. $\mathcal{M} = 138.1$ g/mol

$q = 138.1$ g x $\dfrac{5.80°C \text{ x } 9.37 \text{ kJ}/2.48°C}{1.00 \text{ g}} = 3.03$ x 10^3 kJ

15. a. $Hg(s) \longrightarrow Hg(l)$; $\Delta H = 2.33$ kJ

b. $Br_2(l) \longrightarrow Br_2(g)$; $\Delta H = 29.6$ kJ

c. $C_{10}H_8(l) \longrightarrow C_{10}H_8(s)$; $\Delta H = -19.3$ kJ

17. a. $CS_2(l) + 3 O_2(g) \longrightarrow CO_2(g) + 2SO_2(g)$; $\Delta H = -1075$ kJ

b. exothermic

c. products below reactants

d. $\Delta H = 5.00$ g x $(-1075$ kJ/76.15 g$) = -70.6$ kJ

e. 10.00 kJ x $\dfrac{76.15 \text{ g}}{1075 \text{ kJ}} = 0.7084$ g

19. a. $\Delta H = +81.4$ kJ   b. 2.00 g x $\dfrac{81.4 \text{ kJ}}{111.0 \text{ g}} = 1.47$ kJ

21. a. $C_2H_5SH(g) + 9/2\ O_2(g) \longrightarrow 2CO_2(g) + SO_2(g) + 3H_2O(l)$

$$\Delta H = \frac{-30.21\ kJ}{1.00\ g} \times 62.14\ g = -1.88 \times 10^3\ kJ$$

b. $n\ CO_2 = \dfrac{(10.0\ L)(1.00\ atm)}{(0.0821\ L\cdot atm/mol\cdot K)(298\ K)} = 0.409\ mol$

$$\Delta H = 0.409\ mol \times \frac{-1.88 \times 10^3\ kJ}{2\ mol} = -384\ kJ$$

23. a. $C_{57}H_{104}O_6(s) + 80\ O_2(g) \longrightarrow 57CO_2(g) + 52H_2O(l)$

$$\Delta H = -3.022 \times 10^4\ kJ$$

b. $q\ water = 100\ g \times 4.00°C \times 4.18\ J/g\cdot°C = 1.67 \times 10^3\ J$

$$1.67 \times 10^3\ J \times \frac{885.4\ g}{3.022 \times 10^7\ J} = 4.89 \times 10^{-2}\ g$$

25. $Br_2(l) \longrightarrow Br_2(g);\quad \Delta H = +29.6\ kJ$

$$q = 100.0\ g \times \frac{29.6\ kJ}{159.8\ g} = 18.5\ kJ$$

$H_2O(l) \longrightarrow H_2O(g);\quad \Delta H = 40.7\ kJ$

$$q = 100.0\ g \times \frac{40.7\ kJ}{18.02\ g} = 226\ kJ$$

boiling 100.0 g of water absorbs more heat

27. a. $C_{10}H_8(s) \longrightarrow C_{10}H_8(g)\quad \Delta H = +62.6\ kJ$

b. $Br_2(g) \longrightarrow Br_2(l)\quad \Delta H = -40.4\ kJ$

29. a. $\Delta H_1 = 4.18\ J/g\ °C \times 77.7°C \times 12.50\ g = 4.06 \times 10^3\ J = 4.06\ kJ$

b. $\Delta H_2 = +40.7\ kJ \times \dfrac{12.50}{18.02} = +28.2\ kJ$

c. $\Delta H_3 = 4.06\ kJ + 28.2\ kJ = +32.3\ kJ$

31. $CH_4(g) + 2Cl_2(g) \longrightarrow CH_2Cl_2(g) + 2HCl(g);\quad \Delta H = -202\ kJ$

33. $CH_4(g) \longrightarrow C(s) + 2H_2(g)$ $\Delta H = +74.7$ kJ

$NH_3(g) \longrightarrow \frac{1}{2} N_2(g) + 3/2\ H_2(g)$ $\Delta H = +46.1$ kJ

$\underline{C(s) + \frac{1}{2} H_2(g) + \frac{1}{2} N_2(g) \longrightarrow HCN(g) \qquad \Delta H = +135.2\ kJ}$

$CH_4(g) + NH_3(g) \longrightarrow HCN(g) + 3H_2(g) \qquad \Delta H = +256.0$ kJ

35. a. $C(s) + 2Cl_2(g) \longrightarrow CCl_4(1)$ $\Delta H = -135.4$ kJ

b. $Mg(s) + O_2(g) + H_2(g) \longrightarrow Mg(OH)_2(s)$ $\Delta H = 924.5$ kJ

c. $Ca(s) + S(s) + 2\ O_2(g) \longrightarrow CaSO_4(s)$ $\Delta H = -1434.1$ kJ

d. $\frac{1}{2} H_2(g) + \frac{1}{2} I_2(s) \longrightarrow HI(g)$ $\Delta H = +26.5$ kJ

37. a. -542.2 kJ/2 $= -271.1$ kJ

b. 12.00 g x $\dfrac{-271.1\ kJ}{20.01\ g} = -162.6$ kJ

39. $CaCl_2(s) \longrightarrow Ca^{2+}(aq) + 2Cl^-(aq)$

$\Delta H = -542.8$ kJ $- 334.4$ kJ $+ 795.8$ kJ $= -81.4$ kJ

$q = 5.00$ g x $\dfrac{-81.4\ kJ}{111.0\ g} = -3.67$ kJ

41. a. $\Delta H° = 2\ \Delta H_f° \ Br^- + 2\ \Delta H_f° \ H_2O(1) - 2\ \Delta H_f° \ OH^-(aq)$

$= -243.2$ kJ $- 571.6$ kJ $+ 460.0$ kJ $= -354.8$ kJ

b. $\Delta H° = 3\ \Delta H_f° \ Fe^{3+} + 5\ \Delta H_f° \ OH^- - 1064.0$ kJ $- \Delta H_f° \ CrO_4^{2-}$

$- 4\ \Delta H_f° \ H_2O(1) - 3\ \Delta H_f° \ Fe^{2+}$

$= -145.5$ kJ $- 1150.0$ kJ $- 1064.0$ kJ $+ 881.2$ kJ $+ 1143.2$ kJ

$+ 267.3$ kJ $= -67.8$ kJ

c. $\Delta H° = 3\ \Delta H_f° \ Mn^{2+} + 2\ \Delta H_f° \ NO_3^- + 2\ \Delta H_f° \ H_2O(1) - 3\ \Delta H_f° \ MnO_2(s)$

$- 2\ \Delta H_f° \ NO(g)$

$= -662.4$ kJ $- 410.0$ kJ $- 571.6$ kJ $+ 1560.0$ kJ $- 180.4$ kJ

$= -264.4$ kJ

43. a. $N_2(g) + 3H_2O(g) \longrightarrow 2NH_3(g) + 3/2\ O_2(g)$

$\Delta H° = 2\ \Delta H_f° \ NH_3 - 3\ \Delta H_f° \ H_2O(g) = +633.2$ kJ

b. $2CO_2(g) + 3H_2O(g) \longrightarrow C_2H_5OH(1) + 3\ O_2(g)$

$\Delta H° = \Delta H_f° \ C_2H_5OH - 3\Delta H_f° \ H_2O(g) - 2\Delta H_f° \ CO_2 = 1234.7$ kJ

45. a. $C_2H_4(g) + \frac{1}{2} O_2(g) + 2HCl(g) \longrightarrow C_2H_4Cl_2(1) + H_2O(1)$

$\Delta H° = 318.7$ kJ

b. -318.7 kJ $= \Delta H_f° \ C_2H_4Cl_2 - 285.8$ kJ $+ 184.6$ kJ $- 52.3$ kJ

$\Delta H_f° \ C_2H_4Cl_2 = -217.5$ kJ

47. -1854.9 kJ $= 2(-531.0$ kJ$) + \Delta H_f° \ Cr_2O_3 - 1143.2$ kJ $+ 1490.3$ kJ

$\Delta H_f° \ Cr_2O_3 = -1140.0$ kJ

49. $C_6H_{12}O_6(s) \longrightarrow 2C_2H_5OH(1) + 2CO_2(g)$

$\Delta H° = 2(-277.7$ kJ$) + 2(-393.5$ kJ$) + 1275.2$ kJ $= -67.2$ kJ

$q = 0.120 \text{ L} \times \dfrac{10^3 \text{ mL}}{1 \text{ L}} \times 0.789 \dfrac{g}{mL} \times \dfrac{-67.2 \text{ kJ}}{92.1 \text{ g}} = -69.1$ kJ

51. piston is higher; work done by the system

53. $W = (1.5 \times 0.50) \text{ L·atm} \times \dfrac{101 \text{ J}}{1 \text{ L·atm}} = 76$ J

55. a. $q = \Delta E - W = -88 \text{ J} - 65 \text{ J} = -153$ J

b. $\Delta E = -42 \text{ J} + 115 \text{ J} = +73$ J

57. a. $+40.7$ kJ

b. $q_v = q_p - \Delta n_g RT = +40.7 \text{ kJ} - 1(8.31 \times 10^{-3} \times 373) \text{kJ} = 37.6$ kJ

c. 3.1 kJ

59. a. $C_3H_8(g) + 5\ O_2(g) \longrightarrow 3CO_2(g) + 4H_2O(1) \quad \Delta H° = -2219.9$ kJ

$\Delta H° = 3(-393.5 \text{ kJ}) + 4(-285.8 \text{ kJ}) + 103.8 \text{ kJ} = -2219.9$ kJ

b. $\Delta E° = \Delta H° - \Delta n_g RT = -2219.9 \text{ kJ} + 3(0.00831)(298) \text{ kJ} = -2212.5$ kJ

$1.00 \text{ g} \times \dfrac{2212.5 \text{ kJ}}{44.09 \text{ g}} = 50.2$ kJ

61. $100 \text{ Ncal} \times \dfrac{1 \text{ hr}}{250 \text{ Ncal}} \times \dfrac{60 \text{ min}}{1 \text{ hr}} = 24$ min

63. $2Al(s) + 3NH_4NO_3(s) \longrightarrow 3N_2(g) + 6H_2O(g) + Al_2O_3(s)$

$\Delta H° = \Delta H_f° \ Al_2O_3 + 6 \Delta H_f° \ H_2O(g) - 3 \Delta H_f° \ NH_4NO_3$

$= -1675.7 \ kJ - 1450.8 \ kJ + 1096.8 \ kJ = -2029.7 \ kJ$

theoretical yield based on Al:

$$5.0 \times 10^3 \ g \ Al \times \frac{84.06 \ g \ N_2}{53.96 \ g \ Al} = 7.78 \times 10^3 \ g \ N_2$$

theoretical yield based on NH_4NO_3:

$$5.00 \times 10^3 \ g \ NH_4NO_3 \times \frac{84.06 \ N_2}{240.3 \ g \ NH_4NO_3} = 1.75 \times 10^3 \ g \ N_2$$

Hence, NH_4NO_3 is limiting

$$\Delta H° = 5.00 \times 10^3 \ g \ NH_4NO_3 \times \frac{-2029.7 \ kJ}{240.3 \ g \ NH_4NO_3} = -4.22 \times 10^4 \ kJ$$

65. q brass $= (1.90 \ cm)^3 \times 8.25 \ \frac{g}{cm^3} \times 0.362 \ \frac{J}{g \cdot °C} \times (t - 95.0°C)$

q water $= 20.0 \ g \times 4.18 \ \frac{J}{g \cdot °C} \times (t - 22.0°C)$

$20.5(t - 95.0) = -83.6 \ (t - 22.0)$

$104.1 \ t = 1839 + 1948; \ t = 36.4°C$

67. $n = (1.00(1.00)/(0.0821)(273) = 0.0446 \ mol$

$C_2H_6(g) + 7/2 \ O_2(g) \longrightarrow 2CO_2(g) + 3H_2O(1) \quad \Delta H = -1559.7 \ kJ$

$C_3H_8(g) + 5 \ O_2(g) \longrightarrow 3CO_2(g) + 4H_2O(1) \quad \Delta H = -2219.9 \ kJ$

Let $x = n \ C_2H_6$:

$x(1559.7) + (0.0446 - x)(2219.9) = 75.65; \quad x = 0.0218$

Mole fraction $C_2H_6 = 0.0218/0.0446 = 0.488$

69. a. $12.5 \ g \times 20.0°C \times 6 \times 0.902 \ \frac{J}{g \cdot °C}$

$+ \ \frac{72.0}{16.0} \ lb \times \frac{453.6 \ g}{1 \ lb} \times 20.0°C \times 4.10 \ \frac{J}{g \cdot °C} = 1.68 \times 10^5 \ J$

b. $168 \text{ kJ} \times \dfrac{18.02 \text{ g}}{6.00 \text{ kJ}} = 505 \text{ g}$

70. room temperature $\approx 25°C$

   take 100 g water: $q = 100 \text{ g} \times 25°C \times 4.18 \text{ J/g } °C = 10,400 \text{ J}$

   mass ice required $= 10.4 \text{ kJ} \times \dfrac{1 \text{ mol}}{6.00 \text{ kJ}} \times \dfrac{18.02 \text{ g}}{1 \text{ mol}} = 31 \text{ g}$

   $V_{water} = 100 \text{ g}/(1.00 \text{ g/cm}^3) = 100 \text{ cm}^3$

   $V_{ice} = 31 \text{ g}/(0.90 \text{ g/cm}^3) = 34 \text{ cm}^3$

   $f = 34/134 = 0.25$

71. a. $\Delta H = \Delta H_f° \text{ Al}_2\text{O}_3(s) - \Delta H_f° \text{ Fe}_2\text{O}_3(s) = -851.5 \text{ kJ}$

   b. $851,500 \text{ J} = 0.77 \dfrac{\text{J}}{\text{g} \cdot °C} \times 101.96 \text{ g} \times \Delta t$
   $$+ 0.45 \dfrac{\text{J}}{\text{g} \cdot °C} \times 11.70 \text{ g} \times \Delta t$$

   Solving, $\Delta t = 6600°C$

   c. yes

72. $q = 22.51 \dfrac{\text{kJ}}{°C} \times 1.67°C$; let x = mass sucrose

   $22.51 \dfrac{\text{kJ}}{°C} \times 1.67°C = \dfrac{5640 \text{ kJ}}{342.3 \text{ g}} (x)$

   $x = 2.28 \text{ g}$;   76.0%

Saunders College Publishing

CHAPTER 9
Liquids and Solids

LECTURE NOTES

This chapter requires more lecture time than most. Many of the
concepts are relatively abstract; most are likely to be new to the student.
You should allow 3 lectures for this chapter. Among the more difficult
topics are the following:

- trends in melting and boiling points. Students have a great
deal of difficulty making comparisons between structurally different
types of substances (e.g., NaCl vs Ar, carbon dioxide vs silicon dioxide). At this point, they should be able to classify a substance as
ionic or molecular, given its formula, but many of them still haven't
learned to do this. Students should be expected to recognize some common
network covalent substances such as C, Si, SiC, and SiO_2.

- the geometry of different types of unit cells. Models using
styrofoam balls are a great help here.

- the idea of vapor pressure. Perhaps surprisingly, students have
less trouble with problems involving the Clausius-Clapeyron equation
than they do with problems that test their understanding of what vapor
pressure really means.

<u>LECTURE 1</u>

I <u>Molecular Substances</u>

 A. <u>General properties</u> Nonconductors of electricity, often insoluble
  in water, low melting, low boiling. Molecules are relatively easy
  to separate from each other because intermolecular forces are weak.

 B. <u>Dispersion forces</u> Result from temporary dipoles formed in adjacent
  molecules. Strength depends upon how readily electrons are dis-
  persed. Increase with molecular size, molar mass. Explains why
  boiling point of molecular substances ordinarily increases with
  molar mass (F_2 < Cl_2 < Br_2 < I_2).

 C. <u>Dipole forces</u> Electrical attractive forces between + end of one polar molecule and − end of adjacent molecule. Compare NO (bp = −151°C) to N_2 (bp = −196°C), O_2(bp = −183°C).

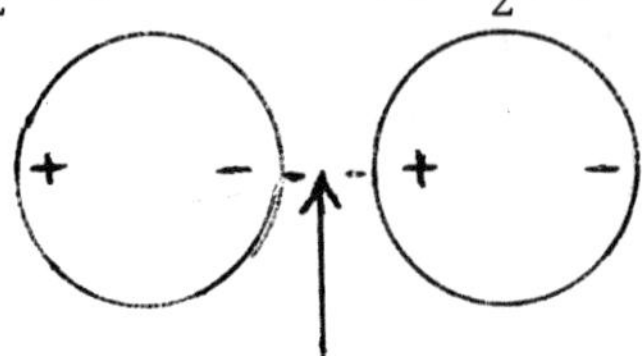

dipole force

D. Hydrogen bonds Unusually strong dipole force. H atom is very small and differs greatly in electronegativity from F, O, or N. Compare boiling points of Group 16 hydrides:

$$H_2O \ (100°C) \qquad H_2S \ (-61°C) \qquad H_2Se \ (-42°C) \qquad H_2Te \ (-2°C)$$

Note that water has many unusual properties in addition to high boiling point. Open structure of ice, a result of hydrogen bonding, accounts for its low density.

II <u>Other Types of Solids</u>

A. <u>Network covalent</u> (C, SiC, SiO_2)

```
  |   |   |
- X - X - X -      High melting (covalent bonds must be broken).
  |   |   |        Nonconducting.  Insoluble in water or other
- X - X - X -      common solvents.  Contrast structure and prop-
  |   |   |        erties of diamond to those of graphite.
- X - X - X -
```

B. <u>Ionic</u> (NaCl, KNO_3)

```
M⁺  X⁻  M⁺      High melting (strong attractive forces between opp-
                ositely charged ions).  Nonconducting as solids;
X⁻  M⁺  X⁻      conduct molten.  Often water soluble; depends on
                balance between attractive forces for each other
M⁺  X⁻  M⁺      and for water molecules.
```

C. <u>Metals</u>

```
M⁺  e⁻  M⁺      "Electron-sea" model; cations in mobile sea of
                electrons.  Conduct electricity.  Ductile, malleable.
e⁻  M⁺  e⁻      Wide range of melting points, depending on number of
                valence electrons.  Insoluble in water.
M⁺  e⁻  M⁺
```

LECTURE 2

I <u>Unit Cells in Metals</u> Unit cell = smallest unit which, repeated over and over again, generates the crystal.

A. <u>Simple cubic</u> Unit cell consists of eight atoms at the corners of a cube.

$$2r = s$$

B. <u>Face-centered cubic</u> Atoms at corners of cube and in center of each face. Atoms touch along face diagonal.

$$4r = s(2)^{\frac{1}{2}}$$

C. <u>Body-centered cubic</u> Atoms at corners of cube and center. Atoms touch along body diagonal.

$$4r = s(3)^{\frac{1}{2}}$$

Sodium crystallizes in BCC structure; unit cell has length of 0.429 nm. Atomic radius?

$$r = s(3)^{\frac{1}{2}}/4 = 0.186 \text{ nm}$$

II Liquid-Vapor Equilibrium

A. <u>Vapor pressure</u> When a liquid is introduced into a closed container, it establishes equilibrium with its vapor:

$$\text{liquid} \rightleftharpoons \text{vapor}$$

The pressure of the vapor at equilibrium is referred to as the vapor pressure of the liquid.

Vapor pressure is independent of volume of container. Add 0.0100 mol of liquid benzene to 1.00 L flask at 25°C (vp benzene = 92 mm Hg). How much benzene vaporizes?

$$n_{vapor} = \frac{PV}{RT} = \frac{(92/760 \text{ atm})(1.00 \text{ L})}{(0.0821 \text{ L·atm/mol·K})(298 \text{ K})} = 0.0050 \text{ mol}$$

$$n_{liquid} = 0.0100 - 0.0050 = 0.0050$$

If the flask were larger than about two liters, all the liquid would vaporize; equilibrium would not be established.

LECTURE 3

I Liquid-Vapor Equilibrium

A. <u>Temperature dependence of vapor pressure</u>

$$\ln P_2/P_1 = \Delta H_{vap}(1/T_1 - 1/T_2)/R$$

$$\Delta H_{vap} = \text{heat of vaporization in joules per mole}$$

87

R is the gas constant, 8.31 J/mol·K

Take the heat of vaporization of benzene to be 30.8 kJ/mol, vp = 92 mm Hg at 25°C. Calculate its vapor pressure at 50°C.

$$\ln P_2/P_1 = 30{,}800(1/298 - 1/323)/8.31 = 0.963$$

$$P_2/P_1 = 2.62; \quad P_2 = 2.62(92 \text{ mm Hg}) = 241 \text{ mm Hg}$$

Note that pressure more than doubles when T rises from 25 to 50°C, reflecting the fact that more benzene vaporizes. Pressure of ideal gas would increase by less than 10%.

B. <u>Boiling point</u> = temperature at which vapor bubbles form in liquid

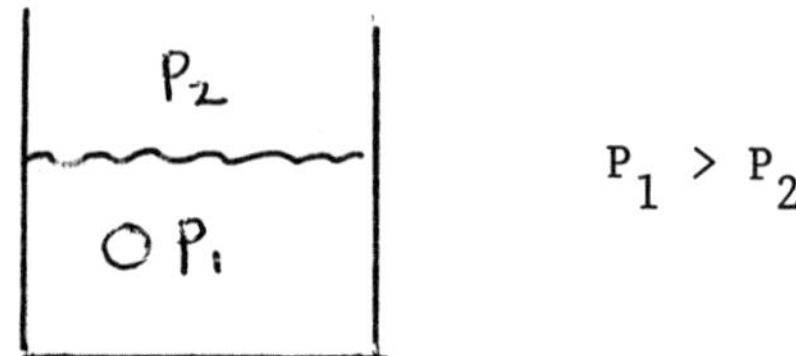

Hence, boiling point varies with applied pressure, P_2. When P_2 = 760 mm Hg, bp water = 100°C. If P_2 = 1075 mm Hg, bp water ▬ 110°C (pressure cooker). If P_2 = 5 mm Hg, water boils at 0°C.

C. <u>Critical temperature</u> Temperature above which liquid cannot exist. Critical pressure = vapor pressure at critical T. Since critical T of oxygen is -119°C, liquid oxygen cannot exist at room temperature, regardless of pressure. Critical T of propane is 97°C; propane is stored as liquid under pressure at room T.

II <u>Phase Diagrams</u>

Graph showing temperatures and pressures at which liquid, solid, and vapor phases of a substance can exist.

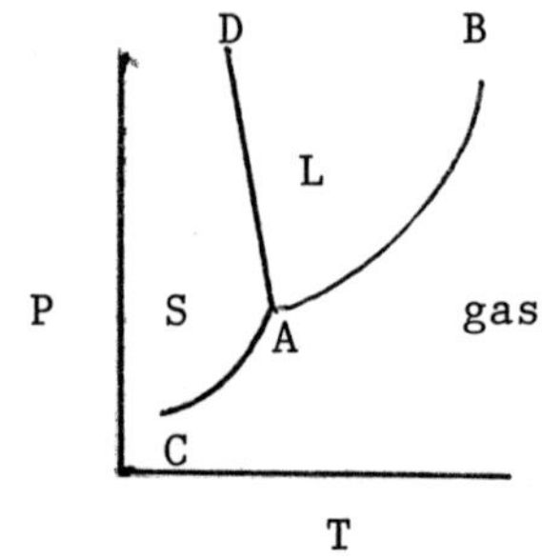

AB = vp curve liquid
AC = vp curve solid
AD = melting point curve
 A = triple point

88

Note that:

    - solid sublimes (passes directly to vapor) below triple point (0°C, 5 mm Hg for water; 115°C, 90 mm Hg for iodine)

    - if line AD slopes toward P axis, melting point decreases as P increases. This behavior is observed for water, where the liquid is the more dense phase. More often, the solid is more dense, AD tilts away from the P axis, and the melting point increases with pressure.

DEMONSTRATIONS

1. Hydrogen bonding in liquids: GILB C 6

2. Conductivity of fused salts: GILB E 17

3. Close packing of spheres: GILB B 19

4. Expansion of water on freezing: SHAK 3 310

5. Ice under pressure: J. Chem. Educ. 67 789 (1990)

6. Liquid-vapor equilibrium: GILB C 1, C 3, D 11; SHAK 1 6

7. Vapor pressure vs temperature: GILB D 13, D 17, D 43; SHAK 2 6, 75, 78

8. Liquefaction of gases: GILB B 4, C 10, D 5, D 12

9. Simultaneous boiling and freezing: J. Chem. Educ. 69 325 (1992); 71 67 (1994)

10. Effect of pressure on boiling point: SHAK 2 81

11. Critical temperature: GILB C 9, R 9; J. Chem. Educ. 69 159 (1992)

<u>VIDEO</u>

1. Expansion of water on freezing: SHAK 37

2. Steam and superheated steam: JCES 24

3. Vapor pressure: SHAK 39

4. Liquefaction of carbon dioxide: SHAK 28; JCES 25

5. Silicon atoms: Falcon 9

PROBLEMS

1. $CH_4 < SiH_4 < GeH_4 < SnH_4$

3. All show dispersion forces; dipole forces in CH_2Cl_2, HCl

5. only CH_3NH_2

7. a. both molecular; dispersion forces stronger in C_8H_{18}

   b. hydrogen bonding in HF

   c. dipole forces in ICl; also, dispersion forces stronger

   d. dispersion forces stronger in O_2

9. c, d

11. a. NO_2; weaker dispersion forces (both are polar)

    b. HCl; NaCl is ionic

    c. Ne; lower dispersion forces

    d. AsH_3; H-bonding in NH_3

13. a. intermolecular forces (dispersion, dipole)

    b. H bonds

    c. intermolecular forces (dispersion, dipole)

    d. ionic bonds

15. a. network covalent   b. ionic   c. metallic

17. a. network covalent, ionic, metallic (usually)

    b. network covalent, metallic, molecular

    c. ionic, metallic

19. a. molecular   b. molecular   c. network covalent

    d. metallic   e. ionic

21. a. Na_2CO_3 c. CH_4 c. steel   d. diamond (C)

23. a. atoms of Si bonded to atoms of O

    b. molecules of SO_2

    c. ions (K^+, O_2^-)

    d. K^+ ions, electrons

25. BCC: $4r = s(1.73)$

27. 0.700 nm $= s(1.414)$; $s = 0.495$ nm

29. a. $s = 2(0.095 + 0.216)$nm $= 0.622$ nm

   b. $1.414(0.622$ nm$) = 0.880$ nm

31. body diagonal $= 2(r \text{ cation} + r \text{ anion}) = s(3)^{\frac{1}{2}}$

   $s = 2(r \text{ cation} + r \text{ anion})/(3)^{\frac{1}{2}} = 1.155(r \text{ cation} + r \text{ anion})$

33. a. 4   b. 8 for Cl^- at corners, 2 for Cl^- in center

35. a. 254 mm Hg x $\dfrac{308 \text{ K}}{330 \text{ K}} = 237$ mm Hg;   254 mm Hg x $\dfrac{318 \text{ K}}{330 \text{ K}} = 245$ mm Hg

   b. 237 mm Hg > 203 mm Hg;   245 mm Hg < 325 mm Hg

   c. 35°C: 203 mm Hg   45°C: 245 mm Hg

   d. 35°C: vapor and liquid   45°C: vapor only

37. a. mass vapor $= \dfrac{(119.4)(186/760)(0.2500)}{(0.0821)(296)}$g $= 0.301$ g   yes

   b. $V = \dfrac{(25.00)(1.498)(0.0821)(296)}{(119.4)(186/760)}$ L $= 31.1$ L

   c. 186 mm Hg x $\dfrac{31.1}{50.0} = 116$ mm Hg

39. a. $\ln \dfrac{760.0}{400.0} = \dfrac{\Delta H(18.5)}{8.31(320)(301)}$; $\Delta H_{vap} = 27.8$ kJ/mol

   b. $\ln \dfrac{P}{400.0} = \dfrac{27,800(12.0)}{(8.31)(313)(301)} = 0.426$; $P = 613$ mm Hg

41. $\ln \dfrac{760.0}{681} = \dfrac{40,700(373 - T)}{(8.31)(373)T} = 0.110$

   $\dfrac{373 - T}{T} = 0.00838$; $T = 370$ K $= 97$°C

43. $\ln \dfrac{1500}{760} = \dfrac{40,700(T - 373)}{(8.31)(373)T} = 0.680$; $\dfrac{T - 373}{T} = 0.0518$

   $T = 393$ K $= 120$°C

45.

ln P	2.30	3.69	4.61	5.99
1/T	3.42×10^{-3}	3.14×10^{-3}	2.95×10^{-3}	2.65×10^{-3}

$$\text{slope} \quad -3.69/(0.77 \times 10^{-3}) = -\Delta H_{vap}/R$$

$$\Delta H_{vap} = 4.0 \times 10^4 \text{ kJ/mol}$$

47. a, b, c

49. a. vapor b. liquid c. solid

51. a. 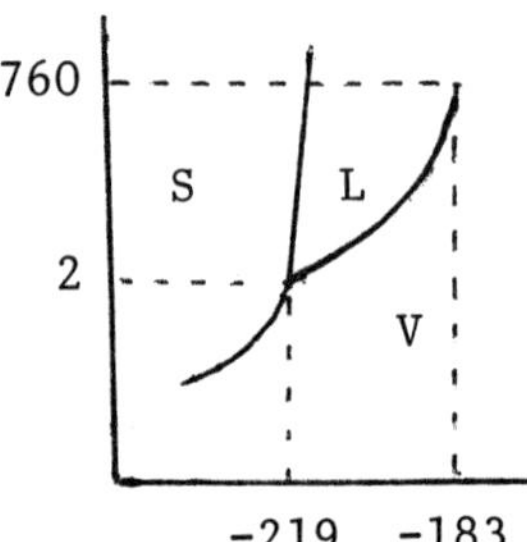 b. yes c. $\sim$ 100 mm Hg

53. a. no change b. liquid $\longrightarrow$ vapor c. vapor $\longrightarrow$ liquid

55. a, b c. liquid $\longrightarrow$ vapor

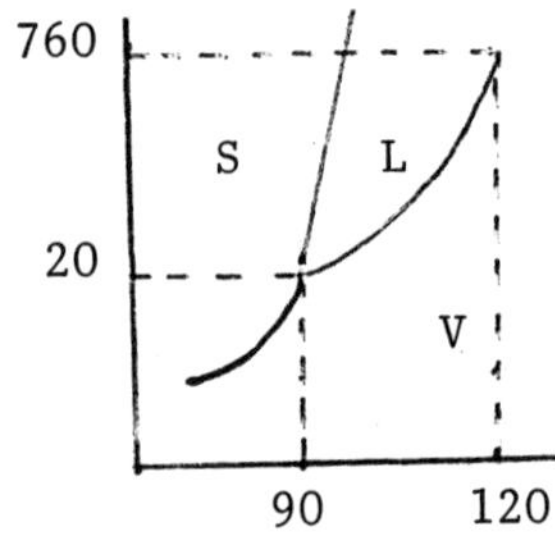

57. a. 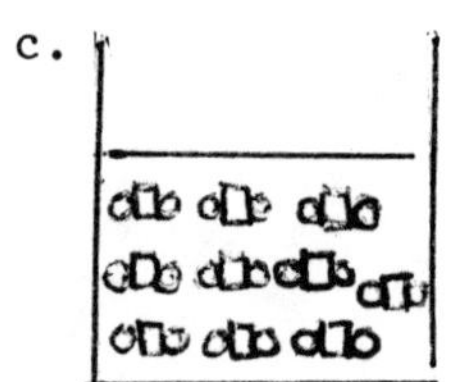 b. 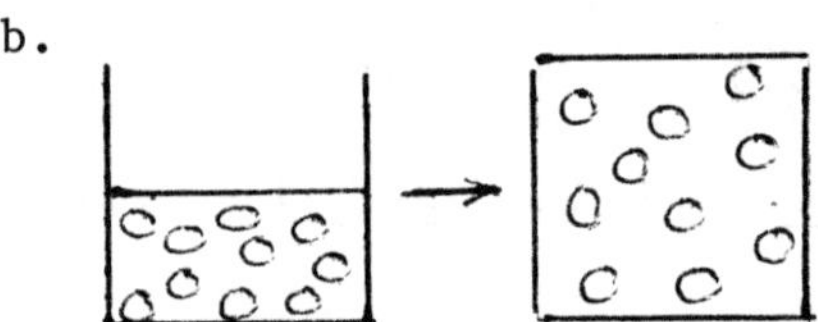

c. 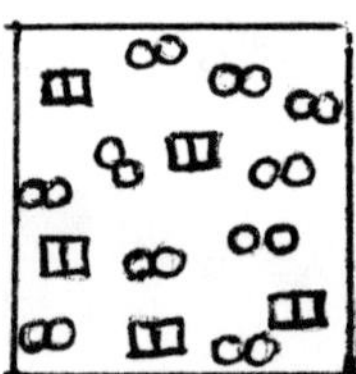

Saunders College Publishing

59. a. true unless V is so large that liquid is completely vaprized

b. false   c. true

d. false; liquid + gas at critical T, P

61. a. all three phases are at equilibrium at the triple point; liquid
and vapor are at equilibrium at critical point

b. metal is made up of + ions and electrons; ionic solid consists
of + and - ions

c. vapor pressure curve is a part of the phase diagram

d. temperature affects vapor pressure, volume doesn't

63. network covalent

65. $P = \dfrac{(2.50/46.03 \text{ mol})(0.0821 \text{ L·atm/mol·K})(293 \text{ K})}{30.0 \text{ L}}$

$= 0.0436$ atm $= 33.1$ mm Hg

66. $V_{molar} = 55.85 \text{ g}/(7.86 \text{ g/cm}^3) = 7.11 \text{ cm}^3$

$0.496 \text{ nm} = s(3)^{\frac{1}{2}}; \quad s = 0.286 \text{ nm}$

$V_{atom} = \dfrac{(0.286 \text{ nm})^3}{2} = 1.17 \times 10^{-2} \text{ nm}^3 = 1.17 \times 10^{-23} \text{ cm}^3$

$N = 7.11/(1.17 \times 10^{-23}) = 6.08 \times 10^{23}$

67. a. 3.60 g water formed

m $H_2O(g) = \dfrac{(26.74/760 \text{ atm})(10.0 \text{ L})(18.02 \text{ g/mol})}{(0.0821 \text{ L·atm/mol·K})(300 \text{ K})} = 0.257$ g

liquid and vapor

b. 26.7 mm Hg

68. $\ln \dfrac{760}{100.0} = \dfrac{\Delta H(54.1)}{8.31 \times 293 \times 347} \qquad \Delta H = 31.7$ kJ/mol

$39°F = 4°C$

$\ln \dfrac{100.0}{P} = \dfrac{31,700 \ (16)}{8.31 \times 293 \times 277} = 0.752; \quad P = 47$ mm Hg

mass vapor $= \dfrac{(133.4)(47/760)(18 \times 28.32)}{(0.0821)(277)}$ g $= 1.8 \times 10^2$ g

68. (cont.)

$$\text{mass orig.} = 0.500 \text{ pt} \times \frac{1 \text{ qt}}{2 \text{ pt}} \times \frac{1 \text{ L}}{1.057 \text{qt}} \times \frac{10^3 \text{ cm}^3}{1 \text{ L}} \times 1.325 \frac{g}{cm^3}$$

$$= 313 \text{ g}$$

$$\% \text{ left} = \frac{133}{313} \times 100 = 42\%$$

69. $\dfrac{120 \text{ lb}}{0.10 \text{ in}^3} \times \dfrac{1 \text{ atm}}{15 \text{ lb/in}^2} = 80 \text{ atm}$

$80/134 = 0.60°C$; heat conduction is more likely

70. $4 \text{ r anion} = (2)^{\frac{1}{2}}(2 \text{ r anion} + 2 \text{ r cation})$

$$\frac{\text{r cation}}{\text{r anion}} = \frac{4 - 2(2)^{\frac{1}{2}}}{2(2)^{\frac{1}{2}}} = 0.4144$$

71. $P \, C_3H_8$ is the vapor pressure of propane, which decreases exponentially with T; $P \, N_2$ is gas pressure which decreases linearly with T

CHAPTER 10
Solutions

LECTURE NOTES

This chapter can be covered in two lectures. The first deals with concentration units, the second with colligative properties. Note that molarity was introduced in Chapter 4. Discussed here for the first time are the calculations involved in preparing a solution of known molarity from a more concentrated solution. The important point to get across is that the number of moles of solute stays the same.

Perhaps the most difficult topic in this chapter is the conversion from one concentration unit to another. Students typically don't know where to start. The material on p. 276 should be helpful here.

<u>LECTURE 1</u>

I <u>Concentrations of Solutes</u>

   A. Mass % solute = $\dfrac{\text{mass solute}}{\text{total mass solution}}$ x 100

     ppm (liquid solution) = mass % x 10^4

     ppb (liquid solution) = mass % x 10^7

   B. Mole fraction: $X_a = \dfrac{\text{no. moles A}}{\text{total number moles}}$

     Dissolve 12.0 g CH_3OH in 100 g water; mole fraction CH_3OH?

     n CH_3OH = 12.0 g x $\dfrac{1 \text{ mol}}{32.0 \text{ g}}$ = 0.375

     n H_2O = 100.0 g x $\dfrac{1 \text{ mol}}{18.02 \text{ g}}$ = 5.55 mol

     X CH_3OH = $\dfrac{0.375}{0.375 + 5.55}$ = 0.0631; X H_2O = 1 − 0.0631 = 0.9369

95

B. <u>Molality</u> $m = \dfrac{\text{no. moles solute}}{\text{no. kg solvent}}$

Molality of solution of CH_3OH referred to above?

$m = 0.375/0.1000 = 3.75$ mol/kg solvent

C. <u>Molarity</u> $M = \dfrac{\text{no. moles solute}}{\text{no. liters solution}}$

1. How prepare 35.0 mL of 0.200 M $Al(NO_3)_3$ from solid?

   molar mass = (26.98 + 42.03 + 144.00)g/mol = 213.01 g/mol

   moles needed = 0.0350 L x 0.200 mol/L = 7.00×10^{-3} mol

   mass needed = 7.00×10^{-3} mol x $\dfrac{213.01 \text{ g}}{1 \text{ mol}}$ = 1.49 g

   Dissolve 1.49 g aluminum nitrate in enough water to form 35.0 mL of solution

2. How prepare 35.0 mL of 0.200 M aluminum nitrate from 0.500 M solution? Note that number of moles of solute stays the same.

   0.0350 L x 0.200 $\dfrac{\text{mol}}{\text{L}}$ = 0.500 $\dfrac{\text{mol}}{\text{L}}$ x V

   V = 0.0140 L = 14.0 mL; dilute to 35.0 mL with water

3. Molality of 0.200 M aluminum nitrate solution (d = 1.012 g/mL)? Work with one liter of solution.

   mass = 1000 mL x $\dfrac{1.012 \text{ g}}{1 \text{ mL}}$ = 1012 g

   mass $Al(NO_3)_3$ = 0.200 mol x $\dfrac{213.01 \text{ g}}{1 \text{ mol}}$ = 42.6 g

   mass water = 1012 g - 43 g = 969 g

   Molality = $\dfrac{0.200 \text{ mol}}{0.969 \text{ kg}}$ = 0.206 mol/kg

II <u>Principles of Solubility</u>

A. <u>Nature of solute and solvent</u> Most nonelectrolytes that are appreciably soluble in water are hydrogen bonded (methanol, hydrogen peroxide, sugars). Other types of nonelectrolytes are generally more soluble in nonpolar or slightly polar solvents such as benzene or toluene.

<u>LECTURE 2</u>

I <u>Principles of Solubility</u>

  A. <u>Effect of temperature</u>

    Increase in T favors endothermic process:

    solid + water $\longrightarrow$ solution ΔH usually positive, so solubility increases with T

    gas + water $\longrightarrow$ solution ΔH usually negative, so solubility decreases with T

  B. <u>Effect of pressure</u>

    Negligible, except for gases, where solubility is directly proportional to the partial pressure of the gas. Carbonated beverages.

II <u>Colligative Properties of Nonelectrolytes</u>
Depend primarily upon concentration of solute particles rather than type.

  A. <u>Vapor pressure lowering</u>

    $VPL = X_2 P_1^\circ$ where X_2 = mole fraction solute, P_1° = vapor pressure pure solvent

  B. <u>Osmosis, osmotic pressure</u>

    Water moves through semi-permeable membrane from region of high vapor pressure (pure water) to region of low vapor pressure (solution)

    π = MRT;   1 M solution at 25°C has osmotic pressure of 24.5 atm

  C. <u>Boiling point elevation, freezing point lowering</u>
    1. Results from vapor pressure lowering.

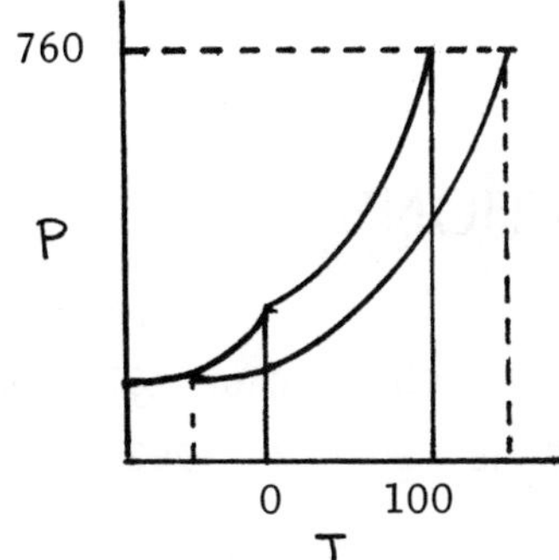

97

2. For water solutions:

$$\Delta T_f = 1.86°C \times m \qquad \Delta T_b = 0.052°C \times m$$

Freezing point and boiling point of solution containing 20.0 g of ethylene glycol (molar mass = 62.0 g/mol) in 50.0 g water?

$$m = \frac{20.0/62.0}{0.050} = 6.45$$

$$\Delta T_f = 6.45 \times 1.86°C = 12.0°C; \quad T_f = -12.0°C$$

$$\Delta T_b = 6.45 \times 0.52°C = 3.4°C \; ; \quad T_b = 103.4°C$$

3. Use in determining molar mass. Suppose solution is prepared by dissolving 0.100 g of nonelectrolyte in 1.00 g of water; freezing point found to be -1.00°C. Molar mass?

$$\text{molality} = 1.00/1.86 \;\; = \;\; \frac{0.100/\mathcal{M}}{0.00100}$$

$$\mathcal{M} = 0.186/0.00100 = 186 \text{ g/mol}$$

III Electrolytes

Colligative effects are greater because of increased number of particles.

$$\Delta T_f = 1.86°C \times m \times i$$

where i is approximately equal to the number of moles of ions per mole of solute:

$$NaCl(s) \longrightarrow Na^+(aq) + Cl^-(aq) \qquad i = 2$$

$$CaCl_2(s) \longrightarrow Ca^{2+}(aq) + 2Cl^-(aq) \qquad i = 3$$

Actually, i is usually less than predicted because of ion atmosphere effects.

DEMONSTRATIONS

1. Supersaturation: GILB F 11; SHAK <u>1</u> 27; J. Chem. Educ. <u>57</u> 152 (1980)

2. Gas solubility: SHAK <u>2</u> 205; J. Chem. Educ. <u>69</u> 573 (1992)

Saunders College Publishing

3. Effect of temperature upon solubility: GILB F 10; SHAK 3 280

4. Raoult's law: GILB F 37, F 38; SHAK 3 242, 254

5. Osmosis: GILB F 41; SHAK 3 283

6. Osmotic pressure: GILB F 42; SHAK 3 286

7. Boiling point elevation: SHAK 3 297

8. Freezing point lowering: SHAK 3 290; J. Chem. Educ. 68 1038 (1991)

VIDEO

1. Supersaturation: SHAK 41; JCES 6

PROBLEMS

1. a. $\dfrac{15.0(0.792)g}{15.0(0.792)g + 25.0(0.7854)g} \times 100 = \dfrac{11.9\ g}{11.9\ g + 19.6\ g} \times 100 = 37.8$

   b. $\dfrac{25.0}{40.0} \times 100 = 62.5$

   c. $\dfrac{11.9/58.05}{11.9/58.05 + 19.6/60.09} = \dfrac{0.205}{0.205 + 0.326} = 0.386$

3. In 100 g of vinegar, there are 5.00 g CH_3COOH

   $$M = \dfrac{5.00/60.05\ mol}{(100/1.006)(0.00100\ L)} = \dfrac{0.0833\ mol}{0.0994\ L} = 0.838\ mol/L$$

5. $(1.00 \times 2.50 \times 10^{-7})$ g Pb $\times \dfrac{1\ mol\ Pb}{207.2\ g\ Pb} = 1.21 \times 10^{-9}$ mol

7. $\mathcal{M}$ $Na_2Cr_2O_7$ = 262.0 g/mol

   a. $M = \dfrac{12.7/262.0\ mol}{0.225\ L} = 0.215\ mol/L$

   b. mass = (2.87 mol/L)(0.386 L)(262.0 g/mol) = 2.90×10^2 g

   c. $\dfrac{3.88/262.0\ mol}{0.693\ mol/L} = 0.0214\ L = 21.4\ mL$

99

9. $\mathcal{M}$ = 342.3 g/mol

    a. Work with 1000 g of water, 3.894 mol sugar = 1333 g sugar

$$\% \text{ water} = \frac{1000 \text{ g}}{1000 \text{ g} + 1333 \text{g}} \times 100 = 42.86$$

$$\text{ppm solute} = \frac{1333}{2333} \times 10^6 = 5.714 \times 10^5$$

$$\text{X solvent} = \frac{1000/18.02}{1000/18.02 + 3.894} = \frac{55.49}{59.38} = 0.9345$$

    b. Work with 100 g of solution; 12.0 g water, 88.0 g sugar

$$m = \frac{88.0/342.3}{0.0120} - 21.4 \text{ mol/kg water}$$

$$\text{ppm} = \frac{88.0 \times 10^6}{100} = 8.80 \times 10^5$$

$$\text{X}_{\text{water}} = \frac{12.0/18.02}{12.0/18.02 + 88.0/342.3} = \frac{0.666}{0.666 + 0.257} = 0.722$$

    c. 1000 g solution; 2.972 g sugar, 997.0 g water

$$m = \frac{2.972/342.3}{0.9970} = 8.709 \times 10^{-3} \text{ mol/kg water}$$

mass % water = 99.70%

$$\text{X solvent} = \frac{997.0/18.02}{997.0/18.02 + 2.992/342.3} = \frac{55.33}{55.33 + 0.009} = 0.9998$$

    d. 1 mol solution; 0.749 mol water, 0.251 mol sugar

$$m = \frac{0.251}{(0.749)(18.02)/1000} = 18.6 \text{ mol/kg water}$$

$$\text{mass \% solvent} = \frac{0.749 \times 18.02}{0.749 \times 18.02 + 0.251 \times 342.3} \times 100 = 13.6$$

$$\text{ppm solute} = 86.4 \times 10^4 = 8.64 \times 10^5$$

11. a. $\mathcal{M}$ $CuSO_4$ = 159.6 g/mol; mass = 0.375 × 0.295 × 159.6 g = 17.7 g

    Weigh out 17.7 g $CuSO_4$, dilute to 375 mL with water

b. $V(1.00$ mol/L$) = 0.375$ L x 0.295 mol/L $= 0.111$ L

   Dilute 111 mL of 1.00 M solution to 375 mL with water

13. 4 squares, 8 circles; 4 M in Ca^{2+}, 8 M in Cl^-

15. a. M $Al_2(SO_4)_3 = 0.188$ M x $325/750 = 0.0815$ mol/L

   M $Al^{3+} = 0.163$ M ; M $SO_4^{2-} = 0.244$ M

   b. 0.325 x 0.188 x $3 = 0.183$ mol Al^{3+}

17.a.Start with 100 g solution; 71.0 g HNO_3

   $$M = \frac{71.0/63.02 \text{ mol}}{(100/1.418)(0.001L)} = 16.0 \text{ mol/L}$$

   b. $(2.00$ L$)(6.00$ mol/L$) = V(16.0$ mol/L$)$

   $V = 0.750$ L; dilute 750 mL of 16.0 M HNO_3 to 2.00 L

19. $\mathcal{M}$ $FeCl_3 = [55.85 + 3(35.45)]$ g/mol $= 162.2$ g/mol

   a. Work with one liter of solution, containing 0.6203 mol $FeCl_3$

   Mass solution = 1080 g; mass $FeCl_3$ = 100.6 g; mass water = 979 g

   $$m = \frac{0.6203}{0.979} = 0.634 \text{ mol/kg water}$$

   $$\text{mass \%} = \frac{100.6}{1080} \times 100 = 9.315\%$$

   b. Work with 1000 g water; n $FeCl_3$ = 1.982; mass $FeCl_3$ = 321.4 g

   $$M = \frac{1.982}{1321/1.230} \times 1000 = 1.845 \text{ mol/L}$$

   $$\text{mass \%} = \frac{321.4}{1321.4} \times 100 = 24.32$$

   c. Work with 100 g solution; 34.47 g $FeCl_3$, 65.53 g water

   $$M = \frac{34.47/162.2 \text{ mol}}{(100/1.350) \times 0.00100 \text{ L}} = 2.869 \text{ mol/L}$$

   $$m = \frac{34.47/162.2}{0.06553} = 3.243 \text{ mol/kg water}$$

Saunders College Publishing

21. a. CH_3OH; H bonding   b. LiCl; ionic vs molecular

    c. H_2O_2; H bonding   d. HCl; H^+ can form H bond with H_2O

23. a. $CaCO_3(s) \longrightarrow Ca^{2+}(aq) + CO_3^{2-}(aq)$

    $\Delta H = -542.8 \text{ kJ} - 677.1 \text{ kJ} + 1206.9 \text{ kJ} = -13.0 \text{ kJ}$

    b. Should decrease

25. a. unsaturated   b. supersaturated   c. unsaturated

27. a. greater   b. less   c. greater   d. less

29. a. $0.21 \text{ atm} \times 3.30 \times 10^{-4} \text{ mol/atm} \times 32.00 \text{ g/mol} = 2.2 \times 10^{-3}$ g

    b. $0.21 \text{ atm} \times 2.85 \times 10^{-4} \text{ mol/atm} \times 32.00 \text{ g/mol} = 1.9 \times 10^{-3}$ g

    c. $\dfrac{0.30}{2.22} \times 100 = 14\%$

31. a. $P = 22.38 \text{ mm Hg} \times 0.850 = 19.0 \text{ mm Hg}$

    b. $X_{H_2O} = \dfrac{77.5/18.02}{77.5/18.02 + 22.5/342.3} = \dfrac{4.30}{4.30 + 0.066} = 0.985$

    $P = 22.38 \text{ mm Hg} \times 0.985 = 22.0 \text{ mm Hg}$

    c. $X_{H_2O} = \dfrac{1000/18.02}{1000/18.02 + 1.88} = \dfrac{55.49}{57.37} = 0.9672$

    $P = 22.38 \text{ mm Hg} \times 0.9672 = 21.65 \text{ mm Hg}$

33. Let x = mass of naphthalene added

    $X_{C_6H_6} = \dfrac{70.00}{74.7} = \dfrac{30.00/78.11}{30.00/78.11 + x/128.2}$

    Solving, x = 3.3 g

35. a. $\pi = (0.250)(0.0821)(298) \text{ atm} = 6.12 \text{ atm}$

    b. $\pi = \dfrac{(10.0)(0.0821)(298)}{(0.055)(60.07)} \text{ atm} = 7.34 \text{ atm}$

    c. $M = \dfrac{(1.50/60.07) \text{ mol}}{(100/1.03)(0.00100 \text{L})} = 0.257 \text{ mol/L}$

    $\pi = (0.257)(0.0821)(298) \text{ atm} = 6.29 \text{ atm}$

102

37. π = $\dfrac{(3.0/58.44)}{1.00}$ x 2 x 0.0821 x 298 atm = 2.5 atm

39. a. m = $\dfrac{32.0/76.1}{0.2500}$; fpl = 3.13°C, bpe = 0.87°C

   -3.13°C, 100.87°C

   b. m = $\dfrac{43.0 \times 0.792/32.04}{0.5000}$; fpl = 3.95°C; bpe = 1.1°C

   -3.95°C, 101.1°C

   c. m = $\dfrac{17.6/92.09}{0.0824}$; fpl = 4.31°C; bpe = 1.2°C

   -4.31°C, 101.2°C

41. 112 g $C_2H_6O_2$, 100 g water

   fpl = $\dfrac{(1.86°C)(112/62.07)}{0.100}$ = 33.6°C

   fp = -33.6°C = -28.5°F; yes

43. 5.9°C = $\dfrac{k(13.66/90.08)}{0.115}$ k = 4.5°C/m

45. 6.5 = $\dfrac{20.2(3.16/\mathcal{M})}{(75.0 \times 0.779)/1000}$ $\mathcal{M}$ = 168 g/mol

   n C = 168 x 0.429/12.0 = 6   n N = 168 x 0.166/14.0 = 2

   n H = 168 x 0.024/1.008 = 4   n O = 168 x 0.381/16.0 = 4

$$C_6H_4N_2O_4$$

47. 0.376 = $\dfrac{20.2 \times 0.0500/\mathcal{M}}{0.00500}$; $\mathcal{M}$ = 537 g/mol

   n C = 0.895 x 537/12.0 = 40   n H = 0.105 x 537/1.008 = 56

   Molecular formula: $C_{40}H_{56}$ Simplest formula: C_5H_7

49. 7.7 = M(0.0821)(298) ; M = 0.32 mol/L

51. $\dfrac{4.60}{760}$ = $\dfrac{3.27 \times 0.0821 \times 293}{\mathcal{M} \times 0.200}$ $\mathcal{M}$ = 6.50 x 10^4 g/mol

53. decreasing fp: $C_2H_6O_2 > K_2Cr_2O_7 > Fe(NO_3)_3 > Cr_2(SO_4)_3$

decreasing bp: $Cr_2(SO_4)_3 > Fe(NO_3)_3 > K_2Cr_2O_7 > C_2H_6O_2$

55. $0.38 = (1.86)(0.20)i$; $i \approx 1$; nonionized

57. One kilogram of maple syrup contains 660 g sucrose, 340 g water

$$m \approx \frac{660/342 \text{ mol}}{20.0 \text{ kg water}} = 0.096 \text{ mol/kg water}$$

(20.0 L of dilute sap solution contains about 20.0 kg water)

59. a. 3 moles of ions in $CaCl_2$ vs 2 in $CaSO_4$

b. solution process is usually endothermic

c. must exceed osmotic pressure of solution

61. a. presence of dissolved salts lowers freezing point

b. vpl is directly proportional to solute mole fraction

c. sugar crystallizes out

d. gas bubbles come out of solution in blood when pressure drops

63. a. If solubility is very low, saturated solution contains little solute

b. sometimes it increases

c. Not even close in concentrated solution

d. approximately 3/2 as great

e. 0.10 m NaCl approximately twice as great

65. 40.0 g glycerol, 60.0 g water

a. $M = \dfrac{40.0/92.09}{(100/1.103)/1000} = 4.79 \text{ mol/L}$

b. $m = \dfrac{40.0/92.09}{0.0600} = 7.24 \text{ mol/kg}$

c. $X \text{ glycerol} = \dfrac{40.0/92.09}{40.0/92.09 + 60.0/18.02} = 0.115$

$P = 12.81 \text{ mm Hg} \times 0.885 = 11.3 \text{ mm Hg}$

d. $\Delta T_f = (1.86°C)(7.24) = 13.5°C$; $T_f = -13.5°C$

67. $m = 3.0/1.04 = 2.9$

$$M = \frac{122}{0.0821 \times 298 \times 2} = 2.49$$

Substituting in the equation in Problem 69:

$2.9 = \dfrac{2.49}{d - 0.243}$; $d = 1.1$ g/mL

68. n KOH = 0.2819 mol

no. kg water $= \dfrac{0.2819}{0.250} = 1.128$; 1128 g water

mass water available = 113 g – 15.82 g = 97 g

Add 1030 g water

69. Consider one liter of solution containing n moles of solute, having density d.

molarity = n; molality $= \dfrac{n}{1000d - n\mathcal{M}} = \dfrac{\text{molarity}}{d - \dfrac{\mathcal{M}\,\text{molarity}}{1000}}$

In dilute solution, d $\longrightarrow$ 1.00 g/mL; molarity $\longrightarrow$ 0, m $\longrightarrow$ M

70. Let x = mass of X

$$\frac{\dfrac{x}{410} + \dfrac{0.100 - x}{342}}{0.00100} = \frac{0.500}{1.86}$$

x = 0.49 g; 49% X

71. $\dfrac{142 \text{ g} \times 0.30 \times 0.15 \times 2}{7.0 \times 10^3 \text{ cm}^3} = 1.8 \times 10^{-3}$ g/cm^3

72. a. Molarity NaOH $= \dfrac{49.92 \text{ g}}{0.600 \text{ L}} \times \dfrac{1 \text{ mol}}{40.0 \text{ g}} = 2.08$ M

b. If NaOH is limiting:

n H$_2$ = 1.248 mol NaOH $\times$ 3/2 = 1.872 mol H$_2$

72. (cont.)

If Al is limiting:

$$n\ H_2 = 41.28\ g\ Al \times \frac{1\ mol\ Al}{26.98\ g\ Al} \times 3/2 = 2.295\ mol\ H_2$$

$$n\ H_2 = 1.872\ mol$$

c. $$V = \frac{(1.872\ mol)(0.0821\ L\cdot atm/mol\cdot K)(298\ K)}{734.8/760\ atm} = 47.4\ L$$

73. $V = nRT/P$ Henry's law: n/P = constant

V = constant x RT

CHAPTER 11
Rate of Reaction

LECTURE NOTES

This chapter is difficult for most students. Concepts such as the rate constant, rate expression, and reaction order are abstract; students have trouble relating chemical kinetics to the real world. Perhaps most difficult of all is the relation between reaction mechanism and order (Section 11.7). Here, students must distinguish carefully between

- the equation for one step in a mechanism and the equation for the overall reaction.

- unstable intermediates and major species (reactants, products)

You should expect to spend at least $2\frac{1}{2}$ lectures on this chapter. If your schedule permits, this could profitably be stretched to 3 lectures.

LECTURE 1

I **Reaction Rate**

  A. **Meaning**   rate = Δ conc. of species/Δt

$$aA + bB \longrightarrow cC + dD$$

$$\text{rate} = \frac{\Delta[C]}{c\Delta t} = \frac{\Delta[D]}{d\Delta t} = \frac{-\Delta[A]}{a\Delta t} = \frac{-\Delta[B]}{b\Delta t}$$

Minus sign used because concentrations of reactants decrease with time. Concentration will be expressed in molarity; time may be in seconds, minutes, years, etc.

  B. **Dependence on concentration of reactant(s)**

Single reactant, A: rate = $k[A]^m$; k = rate constant, m = order

Two reactants, A, B; rate = $k[A]^m[B]^n$ overall order = m + n

107

Saunders College Publishing

Generally, m and n are positive integers (1, 2, 3). However, they can be 0 or a fraction such as ½.

Determination of m and k from rate-concentration data:

$$CH_3CHO(g) \longrightarrow CH_4(g) + CO(g)$$

rate	2.0 M/s	0.50 M/s	0.080 M/s
$[CH_3CHO]$	1.0 M	0.50 M	0.20 M

$$2.0/0.50 = (1.0/0.50)^m \qquad m = 2 \qquad \text{hence, rate} = k[CH_3CHO]^2$$

$$k = \text{rate}/[CH_3CHO]^2 = \frac{2.0 \text{ M/s}}{(1.0 \text{ M})^2} = 2.0(M \cdot s)^{-1}$$

C. <u>First order reaction</u>

$$\ln [A]_o/[A] = kt; \quad [A]_o = \text{original concentration of reactant A}$$

$$[A] = \text{concentration of reactant A at time t}$$

Suppose k = 0.250/s, $[A]_o$ = 1.00 M. What is the concentration of A after 10.0 s?

$$\ln [A]_o/[A] = 0.250 \times 10.0 = 2.50; \quad [A]_o/[A] = e^{2.50} = 12.2$$

$$[A] = 1.00 \text{ M}/12.2 = 0.0819 \text{ M}$$

How long does it take for the concentration to drop to one half its original value?

$$[A] = [A]_o/2; \quad [A]_o/[A] = 2$$

$$\ln 2 = kt; \quad t_{\frac{1}{2}} = \ln 2/k = 0.693/k = 2.77 \text{ s}$$

Note that, for a first order reaction:

- $t_{\frac{1}{2}}$ is independent of original concentration. It takes as long for the concentration to drop from 1.0 M to 0.50 M as it does to drop from 2.0 M to 1.0 M.

- $t_{\frac{1}{2}}$ is inversely related to k. If $t_{\frac{1}{2}}$ is small, k is large and vice versa.

<u>LECTURE 2</u>

I <u>Reactions of Other Orders</u>

Zero order; rate = k; $[A] = [A]_o - kt$; plot of [A] vs t is linear

108

Saunders College Publishing

Second order: rate = $k[A]^2$; $1/[A] - 1/[A]_o = kt$

plot of $1/[A]$ vs t is linear

II Models for Reaction Rate

A. <u>Collision model</u>: rate = $p \times Z \times f$, where $f = e^{-E_a/RT}$. Colliding molecules must be properly oriented (p) and must have certain minimum energy (E_a).

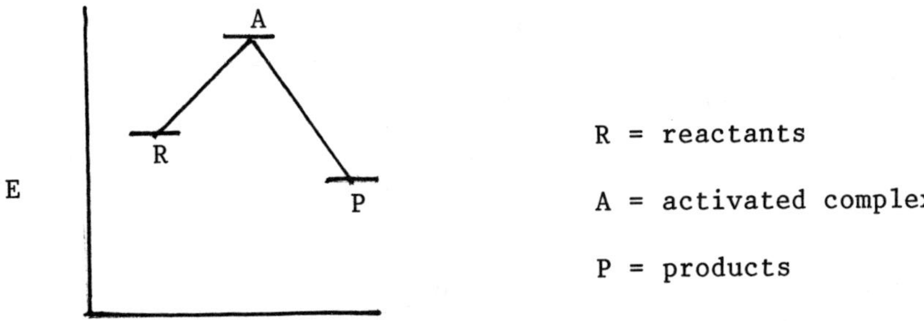

R = reactants

A = activated complex

P = products

B. <u>Activated complex</u>

High energy species in equilibrium with reactants and products.

III <u>Catalysis</u>

Catalyst increases reaction rate without being consumed in the reaction. This happens because catalyst furnishes an alternative path of lower activation energy.

Example: $2H_2O_2(aq) \longrightarrow 2H_2O + O_2(g)$

Direct reaction comes about by collision between H_2O_2 molecules; E_a is high. Reaction is catalyzed by I^- ions:

$$H_2O_2(aq) + I^-(aq) \longrightarrow H_2O + IO^-(aq)$$

$$H_2O_2(aq) + IO^-(aq) \longrightarrow H_2O + O_2(g) + I^-(aq)$$

$$\overline{\quad 2H_2O_2(aq) \longrightarrow 2H_2O + O_2(g) \quad}$$

E_a is much lower for this two-step process.

IV <u>Effect of Temperature</u>

In general, an increase in temperature increases the reaction rate. Rate is approximately doubled when the temperature increases by 10°C. At higher temperature, a greater fraction of the molecules have the activation energy:

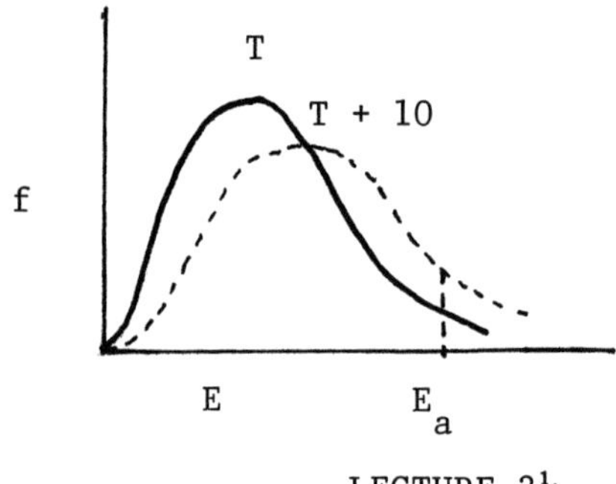

<u>LECTURE 2½</u>

I <u>Temperature and Reaction Rate</u>

  A. <u>Arrhenius Equation</u>

   $\ln k = A - E_a/RT$; $R = 8.31$ J/K, E_a in joules

   Plot of $\ln k$ vs $1/T$ is a straight line with slope $-E_a/R$

   Two point equation:

   $$\ln k_2/k_1 = E_a[1/T_1 - 1/T_2]/R$$

   Suppose rate doubles when T increases from 25 to 35°C. E_a?

   $$\ln k_2/k_1 = 0.693 = E_a[1/298 - 1/308]/8.31$$
   $$E_a = 5.3 \times 10^4 \text{ J} = 53 \text{ kJ}$$

II <u>Reaction Mechanism</u>

  Most reactions take place in a series of steps. To find the rate expression for a multi-step mechanism:

  1 Focus on the slow step; rate for that step = overall rate

  2. In rate-determining step, coefficients give order.

  3. Eliminate any unstable intermediates from rate expression

  Suppose reaction between X_2 and A_2 occurs via a three-step path. First, X_2 undergoes equilibrium dissociation:

   $$X_2(g) \rightleftharpoons 2X(g) \quad \text{fast; rate constants} = k_1(\text{forward}), k_{-1} \text{ (reverse)}$$

  Then:
   $$X(g) + A_2(g) \longrightarrow AX(g) + A(g) \quad \text{slow; rate constant} = k_2$$
   $$A(g) + X_2(g) \longrightarrow AX(g) + X(g)$$

110

rate of reaction = rate slow step = $k_2[X] \times [A_2]$

To eliminate X, an unstable intermediate, note that:

$$k_1[X_2] = k_{-1}[X]^2$$

Solving this equation for [X] and substituting in the rate expression:

$$\text{rate} = \frac{k_2(k_1)^{\frac{1}{2}}[X_2]^{\frac{1}{2}} \times [A_2]}{(k_{-1})^{\frac{1}{2}}} \qquad \text{reaction should be 1st order in } A_2, \tfrac{1}{2} \text{ in } X_2$$

DEMONSTRATIONS

1. Change in concentration with time: GILB I 6
2. Effect of concentration on rate: J. Chem. Educ. _42_ A 607 (1965)
3. Effect of temperature: GILB I 8
4. Catalysis: GILB I 24, I 26; J. Chem. Educ. _70_ 1029 (1993)
5. Catalytic oxidation: GILB I 27, I 31
6. Kinetics of enzyme reactions: GILB I 16
7. Clock reactions: GILB I 42, I 50, I 53; SHAK _4_ 3-86; J. Chem. Educ. _57_ 152 (1980), _69_ 236 (1992)
8. Oscillating reactions: GILB I 55, I 60; SHAK _2_ 248-307
9. Reaction of hydrogen with chlorine: SHAK _1_ 121
10. Reduction of permanganate: J. Chem. Educ. _67_ 598 (1990)
11. Peroxide rocket: GILB I 34
12. UV absorption by ozone: GILB M 198

VIDEO

1. Rates of combustion: Falcon 25
2. Catalysis: SHAK 48
3. Clock reactions: SHAK 44
4. UV absorption by ozone: JCES 33

PROBLEMS

1. a. rate = $\dfrac{\Delta[\text{HOF}]}{2\Delta t}$ b. rate = $-\dfrac{\Delta[\text{O}_2]}{\Delta t}$

3. CO_2: 0.350 mol/L·s x 6 = 2.10 mol/L·s

   O_2: -0.350 mol/L·s x 9 = -3.15 mol/L·s

5. a. $N_2(g) + 3H_2(g) \longrightarrow 2NH_3(g)$

   b. rate = $\Delta[NH_3]/2\Delta t$

   c. rate = $\dfrac{0.448 \text{ mol/L}}{2(12.0) \text{ min}}$ = 0.0187 mol/L·min

7. approximately 0.012 mol/L·s

9. a. 1st order in A, 2nd order in B, 3rd order overall

   b. 3rd order in B, 3rd order overall

   c. 1st order in A, 1st order overall

   d. zero order

11. a. mol/L·min = $k(\text{mol/L})^3$; L^2/mol^2·min

    b. L^2/mol^2·min   c. min^{-1} d. mol/L·min

13. a. (0.830)(0.236) = 0.196 mol/L·hr

    b. k = rate/[A] = 6.41/0.187 = 34.3 hr^{-1}

    c. [A] = rate/k = $(4.99 \times 10^{-2})/0.0591$ = 0.844 mol/L

15. a. rate = $k[\text{Br}_2] \times [\text{NO}]^2$

    b. rate = $1.33 \times 10^{-3} \dfrac{L^2}{\text{mol}^2 \cdot \text{min}} (0.158 \text{ mol/L})^2 (0.316 \text{ mol/L})$

       = 1.05×10^{-5} mol/L·min

    c. $\dfrac{2.8 \times 10^{-8} \text{ mol/L·min}}{(1.33 \times 10^{-3} \, L^2/\text{mol}^2 \cdot \text{min})(0.039 \text{ mol/L})^2}$ = 1.4×10^{-2} mol/L

17. a. k = $\dfrac{0.109 \text{ mol/L·hr}}{(0.450 \text{ mol/L})(0.225 \text{ mol/L})}$ = 1.08 L/mol·hr

    b. $\dfrac{0.222 \text{ mol/L·hr}}{(1.08 \text{ L/mol·hr})(0.155 \text{ mol/L})}$ = 1.33 mol/L

112

c. $\dfrac{0.633 \text{ mol/L} \cdot \text{hr}}{(1.08 \text{ L/mol} \cdot \text{hr})(1/3)} = [NO_2^-]^2; \quad [NO_2^-] = 1.33 \text{ mol/L}$

19. d

21. a. $\dfrac{0.0380}{0.0260} = (0.0731/0.0500)^n; \quad 1.46 = (1.46)^n;$ 1st order

    b. rate $= k[CH_3COCH_3]$

    c. $\dfrac{0.0341}{0.0655} = 0.521/\text{min}$

23. a. $\dfrac{1.82 \times 10^{-2}}{6.00 \times 10^{-4}} = (5.5)^n = 30.3; \ n = 2$ 2nd order in ClO_2

       $\dfrac{1.50 \times 10^{-3}}{6.00 \times 10^{-4}} = (2.5)^m = 2.5; \ m = 1$ 1st order in OH^-
                                                            3rd order overall

    b. rate $= k[OH^-][ClO_2]^2$

    c. $k = \dfrac{6.00 \times 10^{-4}}{(0.030)(0.010)^2} = \dfrac{6.00 \times 10^{-4}}{3.0 \times 10^{-6}} = 2.0 \times 10^2 \ L^2/\text{mol}^2 \cdot s$

    d. rate $= 2.0 \times 10^2 (0.25)^2 (0.036) = 0.45 \text{ mol/L} \cdot s$

25. a. $\dfrac{1.78 \times 10^{-4}}{8.89 \times 10^{-5}} = 2^n; \ n = 1$ 1st order in I^-

       $\dfrac{1.78 \times 10^{-4}}{8.89 \times 10^{-5}} = 2^m; \ m = 1$ 1st order in BrO_3^-

       $\dfrac{3.56 \times 10^{-4}}{8.89 \times 10^{-5}} = 2^p; \ p = 2$ 2nd order in H^+

    b. rate $= k[I^-][BrO_3^-][H^+]^2$

    c. $k = \dfrac{8.89 \times 10^{-5}}{(2.0 \times 10^{-3})(8.0 \times 10^{-3})(2.0 \times 10^{-2})^2} = 1.4 \times 10^4 \ \dfrac{L^3}{\text{mol}^3 \cdot s}$

    d. $\dfrac{5.00 \times 10^{-4}}{(0.0075)(0.0150)(1.4 \times 10^4)} = [H^+]^2; \quad [H^+] = 0.018 \text{ M}$

27. a. 1st order in SCN^-, 1st order in $Cr(H_2O)_6^{3+}$

       rate $= k[Cr(H_2O)_6^{3+}][SCN^-]$

b. $\dfrac{6.5 \times 10^{-4}}{(0.025)(0.060)} = 0.43$ L/mol·min

c. rate $= \dfrac{(0.43)(0.0500)(0.015/97.2)}{1.50} = 2.2 \times 10^{-6}$ mol/L·min

29.

t	0	20	60	120	160	
x	0.368	0.333	0.287	0.235	0.208	
1/x	2.72	3.00	3.48	4.26	4.81	linear
ln x	−1.00	−1.10	−1.25	−1.45	−1.57	

2nd order: rate $= k[C_{12}H_{22}O_{11}]^2$

31. Half-life dircetly proportional to original concentration; zero order

33. a.

t	0	1	2	3	4	8	16
ln [ ]	−2.30	−2.33	−2.35	−2.38	−2.40	−2.50	−2.70

linear plot

b. $k \approx 0.40/16 = 0.025$/min

c. $\ln \dfrac{0.100}{0.0400} = 0.025\ t$; 37 min

d. $(0.025)(0.200) = 0.0050$ mol/L·min

35. a. $\ln \dfrac{0.0100}{x} = \dfrac{0.125}{12} = 0.0104;\ x = 0.0099$ mol/L

b. $\ln \dfrac{0.0100}{0.0075} = 0.125\ t;\ t = 2.3$ yr

c. $\dfrac{0.693}{0.125/\text{yr}} = 5.54$ yr

37. a. $\ln \dfrac{0.250}{0.117} = k(13.7\ \text{hr}) = 0.759;\ k = 0.0554\ \text{hr}^{-1}$

b. $t_{\frac{1}{2}} = 0.693/(0.0554\ \text{hr}^{-1}) = 12.5$ hr

c. rate $= 0.0554$ mol/L·hr

d. $\ln \dfrac{1.00}{0.050} = 0.0554\ t = 3.00;\ t = 54$ hr

39. $\ln 10.0/x = 0.0541\ \text{hr}^{-1} \times 24\ \text{hr} = 1.30;\ x = 2.73$ mg

41. $\ln \dfrac{0.050}{0.016} = k(24.9 \text{ hr})$; $k = 0.046 \text{ hr}^{-1}$

$t_{\frac{1}{2}} = \dfrac{0.693 \text{ hr}}{0.046} = 15 \text{ hr}$

43. a. $\dfrac{1}{0.128} - \dfrac{1}{0.200} = 0.84t$; $t = 3.3 \text{ min}$

b. $t_{\frac{1}{2}} = \dfrac{1}{(0.84)(0.100)} = 12 \text{ min}$

c. rate $= (0.84)(0.078)^2 = 5.1 \times 10^{-3} \text{ mol/L}\cdot\text{min}$

45. a. $\dfrac{0.750}{2(0.0125)} = 30.0 \text{ min}$

b. $\dfrac{0.240 \text{ mol/L}}{0.0125 \text{ mol/L}\cdot\text{min}} = 19.2 \text{ min}$

47. a. A   b. C

49. 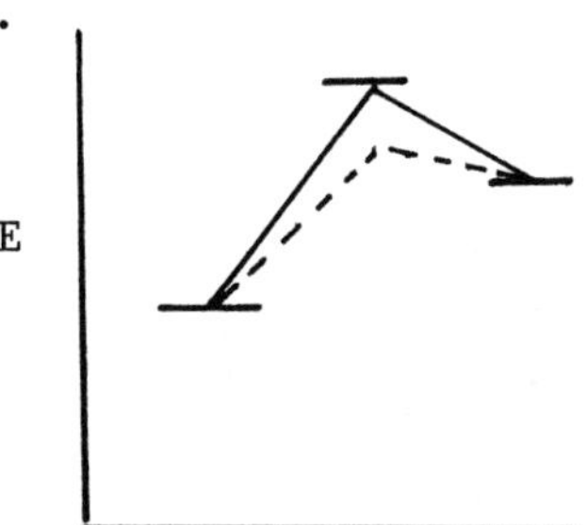

51.
$\ln k$	0.45	1.28	2.05	2.76
$1/T$	0.00161	0.00154	0.00149	0.00143

$E_a \approx \dfrac{8.31 \times 2.31}{0.00018} = 1.1 \times 10^2 \text{ kJ}$

53. a. $0.693/(4.90 \times 10^{-4}) = 1.41 \times 10^3 \text{ min}$

b. $\ln t_2/t_1 = \dfrac{420,000(3)}{8.31 \times 313 \times 310}$; $t_2/t_1 \approx 5$; $t_2 \approx 3 \times 10^2 \text{ min}$

c. 5 per 3°C; a factor of about 2 per °C

55. $\ln k_2/k_1 = \dfrac{98{,}000(5.0)}{8.31(310)(305)}$; $k_2/k_1 = 1.9$

$k_1 = 0.54k_2$; drops by 46%

57. a. $\ln 2 = \dfrac{181{,}000(1273 - T)}{8.31(1273)\,T}$

$\dfrac{1273 - T}{T} = 0.0405$; $T = 1223$ K $= 950°C$

b. $\ln \dfrac{1.6 \times 10^3}{k} = \dfrac{181{,}000(275)}{(8.31)(1273)(998)}$; $\dfrac{1.6 \times 10^3}{k} = 112$

$k = 14$ L/mol·s

59. a. rate $= k[NO_3][CO]$

b. rate $= k[NO][O_2]$

c. rate $= k[NO_2]^2$

61. rate $= k[N_2O_2][H_2]$

$k_1[NO]^2 = k_2[N_2O_2]$

rate $= \dfrac{k \cdot k_1[NO]^2[H_2]}{k_2}$ yes

63. c, d

65. $\ln 10/6 = k(2)$; $k = 0.26$

3 min; $\ln 10/x = 0.78$; $x = 5$; 5 circles, 5 squares, 5 triangles

4 min: $\ln 10/x = 1.04$; $x = 4$; 4 circles, 6 squares, 6 triangles

$t_{\frac{1}{2}} = 0.693/0.26 \approx 3$ min

67. Expt 1 : rate $= 0.5[0.100] = 0.05$ mol/L·min

Expt 2 : $k = 0.5$ min^{-1}; rate $= 0.1$ mol/L·min ; $E_a = 32$ kJ

Expt 3: rate $= 0.1$ mol/L·min; $E_a = ?$

Expt 4: rate $= 0.06$ mol/L·min; $E_a = 32$ kJ

69. a. rate usually decreases (except for zero order)

b. usually increases (increases P and hence concentration)

c. increases

71. $\dfrac{1}{x} - \dfrac{1}{0.100} = 0.80 \times 60$; $\dfrac{1}{x} = 48 + 10 = 58$; $x = 0.017$ mol/L

Δconc. $= 0.083$ mol/L

ΔH per mole $= 90.2$ kJ $+ 15.5$ kJ $- 82.2$ kJ $= 23.5$ kJ

ΔH $= 23.5$ kJ $\times 0.083 = 2.0$ kJ

73. $\ln k_2/k_1 = \dfrac{100,000(15)}{(8.31)(298)(283)} = 2.1$

$k_2/k_1 = 8.5$; 750% increase

75. $-\Delta[A] = k\,\Delta t$; $[A]_o - [A] = kt$

77. a. $\Delta H° = 52.96$ kJ $- 62.44$ kJ $= -9.48$ kJ

$E_{ar} = 165$ kJ $+ 9$ kJ $= 174$ kJ

b. $\dfrac{k_r}{k} = \dfrac{e^{-E_{ar}/RT}}{e^{-E_a/RT}} = e^{(E_a - E_{ar})/RT} = e^{-9/(0.00831 \times 973)} = e^{-1.1}$

$k_r = 0.33\,k = 46$ L/mol$\cdot$s

c. 1.8 mol/L$\cdot$s

78. $\dfrac{-d[A]}{a\,dt} = k[A]$; $\dfrac{-d[A]}{[A]} = ak\,dt$; $\ln [A]_o/[A] = akt$

79. a. $\dfrac{-d[A]}{dt} = k[A]^2$; $\dfrac{-d[A]}{[A]^2} = k\,dt$; $\dfrac{1}{[A]} - \dfrac{1}{[A]_o} = kt$

b. $\dfrac{-d[A]}{[A]^3} = k\,dt$; $\dfrac{1}{2[A]^2} - \dfrac{1}{2[A]_o^2} = kt$

80. Compare 1st and 4th: 2nd order in A

Compare 2nd and 4th: 1st order in C

Compare 1st and 3rd; change in [C] should multiply rate by 2, so must
be first order in B

80. (cont.)

$$\text{rate} = k[A]^2[B][C]$$

81. $a = 0.30 \times \dfrac{8 \text{ hr}}{48 \text{ hr}} = 0.050$; saturation value $= \dfrac{0.100 \text{ g}}{1 - 10^{-0.050}} = 0.91 \text{ g}$

$0.500 = \dfrac{X}{0.11}$ $X = 0.055 \text{ g}$

Saunders College Publishing

CHAPTER 12
Gaseous Chemical Equilibrium

LECTURE NOTES

You will notice that this chapter deals exclusively with the thermodynamic equilibrium constant (K_P) referred to here simply as K. A major advantage of this approach is that, in future chapters, you don't have to hedge on relations such as that between $\Delta G°$ and K, the temperature dependence of K, etc. It is important to point out (p. 348) that in the expression for K, it is always true that

- gases enter as their partial pressures in atmospheres
- aqueous solutes enter as their molar concentrations
- solids, pure liquids, and the solvent do not appear.

Most students find equilibrium calculations difficult to follow, let alone carry out on their own. This is particularly true for the most important calculation: finding equilibrium pressures of all species. Equilibrium tables such as the one shown with Example 12.6 help.

This chapter will require at least 2 lectures, more likely $2\frac{1}{2}$.

<u>LECTURE 1</u>

I <u>The Equilibrium Constant, K</u>

A. $2HI(g) \rightleftharpoons H_2(g) + I_2(g)$

To study this equilibrium at 520°C, put pure HI or a mixture of hydrogen and iodine in a sealed container at this temperature. Take samples over a period of time until the partial pressures become constant. At that point, the system is in equilibrium.

	$2HI(g)$	$H_2(g)$	+	$I_2(g)$	
P_o(atm)	0.200	0.000		0.000	
ΔP(atm)	−0.040	+0.020		+0.020	Expt. 1
P_{eq}(atm)	0.160	0.020		0.020	

119

P_o(atm)	0.100	0.100	0.100	
ΔP(atm)	+0.140	−0.070	−0.070	Expt 2
P_{eq}(atm)	0.240	0.030	0.030	

Note that:

a. $\Delta P\ HI = -2\ \Delta P\ H_2 = -2\ \Delta P\ I_2$

b. $\dfrac{P\ H_2 \times P\ I_2}{(P\ HI)^2} = 0.016$ at equilibrium in all cases

In general, at any temperature, the quotient:

$$\frac{P\ H_2 \times P\ I_2}{(P\ HI)^2}$$

is a constant, equal to K. The value of K depends only upon temperature; it is independent of original pressures, total pressure, or volume.

II General Expression for K

A. Only gases involved

$$aA(g) + bB(g) \rightleftharpoons cC(g) + dD(g)$$

$$K = \frac{(P\ C)^c \times (P\ D)^d}{(P\ A)^a \times (P\ B)^b}$$

Note that products (right side of equation) appear in numerator, reactants in denominator.

B. Solids or liquids as well as gases

$$CaCO_3(s) \rightleftharpoons CaO(s) + CO_2(g) \qquad K = P\ CO_2$$

Terms for solids or liquids do not appear. Adding or removing a solid or liquid does not affect the position of an equilibrium.

C. Aqueous solutions (covered in later chapters in more detail). Solutes appear as molarities; solvent does not appear

$$Ag(s) + 2H^+(aq) + NO_3^-(aq) \rightleftharpoons Ag^+(aq) + NO_2(g) + H_2O$$

$$K = \frac{[Ag^+] \times P\ NO_2}{[H^+]^2 \times [NO_3^-]}$$

$\rlap{-}D$. <u>Relations between equilibrium constants</u>

$$2HI(g) \; \rightleftharpoons \; H_2(g) + I_2(g) \qquad K = 0.016$$

$$H_2(g) + I_2(g) \; \rightleftharpoons \; 2HI(g) \qquad K' = 1/K = 62$$

$$HI(g) \; \rightleftharpoons \; \tfrac{1}{2}H_2(g) + \tfrac{1}{2}I_2(g) \qquad K'' = K^{\frac{1}{2}} = 0.13$$

Rule of Multiple Equilibria:

If Equation 3 = Equation 1 + Equation 2, then $K_3 = K_1 + K_2$

LECTURE 2

I <u>Calculation of K</u>

$$N_2O_4(g) \; \rightleftharpoons \; 2NO_2(g)$$

At 100°C , P N_2O_4 = 0.50 atm, P NO_2 = 2.3 atm at equilibrium

$K = (2.3)^2/0.50 = 11$

II <u>Applications of K</u>

In general, if K is large, equilibrium moves far to right; products favored. If K is small, equilibrium position favors reactants.

A. <u>Determination of direction of reaction</u>

Compare the actual pressure quotient, Q, to that required at equilibrium, K. If Q < K, forward reaction occurs to make Q larger. If Q > K, reverse reaction occurs to make Q smaller. If Q = K, system is already at equilibrium, so nothing happens.

$$2HI(g) \; \rightleftharpoons \; H_2(g) + I_2(g) \qquad K = 0.016$$

Suppose P HI = 0.100 atm, no H_2 or I_2

Q = 0, system shifts to right

Suppose P HI = P H_2 = P I_2 = 0.100 atm

Q = 1, system shifts to left

B. <u>Extent of Reaction</u>

1. Start with pure HI at 0.100 atm. Equilibrium pressures of all species?

121

$$2HI(g) \rightleftharpoons H_2(g) + I_2(g)$$

	2HI(g)	H$_2$(g)	I$_2$(g)
P_o	0.100	0.000	0.000
ΔP	$-2x$	$+x$	$+x$
P_{eq}	$0.100 - 2x$	x	x

$$\frac{x^2}{(0.100 - 2x)^2} = 0.016; \qquad \frac{x}{0.100 - 2x} = 0.13; \qquad x = 0.010$$

$$P\,HI = 0.080 \text{ atm}; \quad P\,H_2 = P\,I_2 = 0.010 \text{ atm}$$

2. Start with pure N_2O_4 at 1.00 atm. Equilibrium partial pressures of N_2O_4, NO_2?

$$N_2O_4(g) \rightleftharpoons 2NO_2(g) \qquad K = 11$$

	N_2O_4(g)	NO_2(g)
P_o	1.00	0.00
P	$-x$	$+2x$
P_{eq}	$1.00 - x$	$2x$

$$4x^2/(0.100 - x) = 11; \quad x = 0.78 \text{ (by quadratic formula)}$$

$$P\,NO_2 = 1.56 \text{ atm}, \quad P\,N_2O_4 = 0.22 \text{ atm}$$

LECTURE 2½

I Effect of Changes in Conditions on Equilibrium Position

Le Chatelier's principle: If a system at equilibrium is disturbed by a change in concentration, pressure, or temperature, the system will, if possible, shift so as to partially counteract the change.

A. Adding or removing a gaseous species

$$2HI(g) \rightleftharpoons H_2(g) + I_2(g)$$

If H_2 is added, part of it reacts; P HI greater than before equilibrium was disturbed, P I_2 less. P H_2 is intermediate between original equilibrium value and that immediately after equilibrium was disturbed.

If H_2 is removed, some HI decomposes to bring P H_2 back part way to its original value. P I_2 increases, P HI decreases.

B. Increase in pressure

Immediate effect of compression is to increase concentration of

122

molecules. To counteract this, reaction occurs which decreases number
of molecules in gas phase.

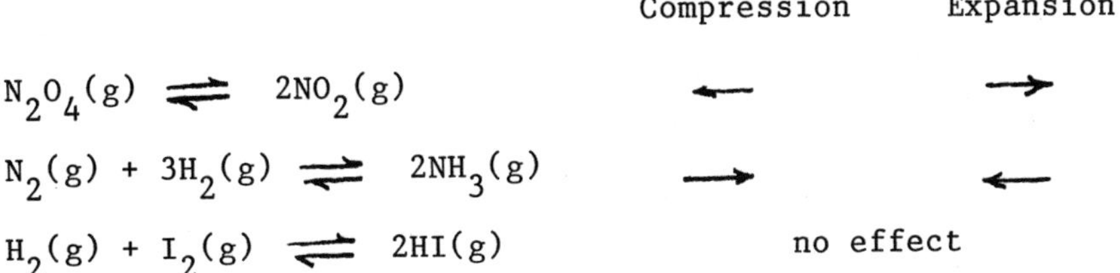

C. <u>Change in temperature</u>

Effect of increase in T is partially counteracted if endothermic
process occurs. Hence, endothermic process is favored by increase in T

	ΔH	increase T	decrease T
$N_2O_4(g) \rightleftharpoons 2NO_2(g)$	+58.2 kJ	→	←
$N_2(g) + 3H_2(g) \rightleftharpoons 2NH_3(g)$	−92.4 kJ	←	→

If forward reaction is endothermic, K increases as T increases; if
exothermic, K decreases as T increases.

DEMONSTRATIONS

1. Mechanical model of equilibrium: J. Chem. Educ. <u>67</u> 598 (1990)
2. N_2O_4 − NO_2 system: GILB J 13; SHAK <u>2</u> 180
3. Le Chatelier's principle: GILB J 10, J 18
4. Catalytic oxidation of ammonia: SHAK <u>2</u> 214

<u>VIDEO</u>

1. Shifting equilibrium; adding or removing a substance: JCES 26
2. Shifting equilibrium; changing volume: JCES 27

PROBLEMS

123

1. a. approximately 80 s   b. faster; same

3.   0   1   2   3   4   5   6

   0.800 0.717 0.633 0.583 0.540 0.533 0.533

   1.000 0.834 0.667 0.568 0.482 0.468 0.468

   0.000 0.249 0.500 0.649 0.778 0.800 0.800

5. a. $\dfrac{(P\ XeF_6)}{(P\ Xe)(P\ F_2)^3}$ b. $\dfrac{(P\ CS_2)(P\ H_2)^4}{(P\ CH_4)(P\ H_2S)^2}$ c. $\dfrac{(P\ C_3H_8)(P\ O_2)^5}{(P\ CO_2)^3(P\ H_2O)^4}$

7. a. $1/(P\ O_2)$ b. $\dfrac{[H^+]\ \times\ [HCOO^-]}{[HCOOH]}$ c. $\dfrac{[Al^{3+}]^2\ \times\ (P\ H_2)^3}{[H^+]^6}$

9. a. $CCl_4(l) \rightleftharpoons CCl_4(g)$ $K = P\ CCl_4$

   b. $3N_2H_4(g) + 4ClF_3(g) \rightleftharpoons 12HF(g) + 3N_2(g) + 2Cl_2(g)$

   $K = \dfrac{(P\ HF)^{12}(P\ N_2)^3(P\ Cl_2)^2}{(P\ N_2H_4)^3(P\ ClF_3)^4}$

   c. $CO_3^{2-}(aq) + 2H^+(aq) \rightleftharpoons CO_2(g) + H_2O$

   $K = \dfrac{(P\ CO_2)}{[CO_3^{2-}]\ \times\ [H^+]^2}$

11. a. $2IF(g) \rightleftharpoons I_2(g) + F_2(g)$

    b. $Cl_2(g) + 2I^-(aq) \rightleftharpoons 2Cl^-(aq) + I_2(s)$

    c. $C_3H_8(g) + 5\ O_2(g) \rightleftharpoons 3CO_2(g) + 4H_2O(g)$

    d. $N_2(g) + 3H_2(g) \rightleftharpoons 2NH_3(g)$

    e. $2H_2O_2(g) \rightleftharpoons 2H_2O(g) + O_2(g)$

13. a. $(0.0125)^{\frac{1}{2}} = 0.112$

    b. $1/(0.0125)^2 = 6.40 \times 10^3$

15. $SnO_2(s) + 2CO(g) \rightleftharpoons Sn(s) + 2CO_2(g)$ $\quad K = 1/(6.45)^2$

$2CO_2(g) + 2H_2(g) \rightleftharpoons 2CO(g) + 2H_2O(g)$ $\quad K = 1/(0.034)^2$

$$SnO_2(s) + 2H_2(g) \rightleftharpoons Sn(s) + 2H_2O(g) \quad K = 1/(6.45 \times 0.034)^2 = 21$$

17. $N_2O_4(g) \rightleftharpoons N_2O(g) + 3/2\,O_2(g)$ $\quad K = 1/(1.2 \times 10^6)^{\frac{1}{2}}$

$N_2O(g) \rightleftharpoons N_2(g) + \frac{1}{2}\,O_2(g)$ $\quad K = 1/(1.2 \times 10^{-35})^{\frac{1}{2}}$

$N_2(g) + 2\,O_2(g) \rightleftharpoons 2NO_2(g)$ $\quad K = 1.7 \times 10^{-17}$

$$N_2O_4(g) \rightleftharpoons 2NO_2(g) \quad K = \frac{(1.7 \times 10^{-17})}{(1.2 \times 10^6 \times 1.2 \times 10^{-35})^{\frac{1}{2}}} = 4.5 \times 10^{-3}$$

19. a. $CS_2(s) + 4H_2(g) \rightleftharpoons CH_4(g) + 2H_2S(g)$

b. $K = \dfrac{(0.131)(0.084)^2}{(0.428)(0.921)^4} = 3.00 \times 10^{-3}$

21. $P(atm) = 0.0821 \times 373 \times molarity = 30.6 \times molarity$

$K = \dfrac{(30.6)(0.0128) \times (30.6)^2(0.0155)^2}{30.6(0.00244)} = 1.18$

23. $P\ NO = (0.981\ atm)(0.254) = 0.249\ atm;$ $\quad \Delta P = -0.732\ atm$

$\Delta P\ Cl_2 = -0.366\ atm;$ $\quad P\ Cl_2 = 0.483\ atm - 0.366\ atm = 0.117\ atm$

$\Delta P\ NOCl = +0.732\ atm;$ $\quad P\ NOCl = 0.100\ atm + 0.732\ atm = 0.832\ atm$

$K = \dfrac{(0.832)^2}{(0.117)(0.249)^2} = 95.4$

25. a. $Q = \dfrac{0.12}{0.28 \times 0.071} = 6.0 > K\ ;$ no

b. $\leftarrow$

27. a. $Q = \dfrac{(0.44)^2(0.44)^{\frac{1}{2}}}{(0.44)(0.00)} \rightarrow \infty\ ;$ $Q > K;$ $\quad \leftarrow$

b. $Q = \dfrac{(0.15)^2(0.22)^{\frac{1}{2}}}{(0.093)(0.27)} = 0.42;$ $Q < K;$ $\quad \rightarrow$

29. b

Saunders College Publishing

31. $K = \dfrac{(P \; COBr_2)}{(P \; CO)(P \; Br_2)}$; $P \; Br_2 = \dfrac{(P \; COBr_2)}{K(P \; CO)} = \dfrac{0.163}{0.185 \times 0.734} = 1.20$ atm

$P_{tot} = 0.163$ atm $+ 0.734$ atm $+ 1.20$ atm $= 2.10$ atm

33. $P \; NO_2 / P \; NO = 1.31(0.643)^{\frac{1}{2}} = 1.05$

$2.05(P \; NO) = 0.42$ atm; $P \; NO = 0.20$ atm; $P \; NO_2 = 0.22$ atm

35. $(2x)^2 / (0.500 - x)^2 = 47$; $2x/(0.500 - x) = 6.9$; $x = 0.385$

$P \; HCN = 0.77$ atm; $P \; C_2N_2 = P \; H_2 = 0.12$ atm

37. a. $\dfrac{(0.159 + x)^2}{(0.485 - x)^2} = 1.30$; $\dfrac{0.159 + x}{0.485 - x} = 1.14$; $x = 0.184$

$P \; H_2 = P \; CO_2 = 0.343$ atm; $P \; CO = P \; H_2O = 0.301$ atm

b. same; no, only if there is no change in number of moles

39. a. $x^2 = 0.215$; $x = 0.464$ atm; $P_{tot} = 0.928$ atm

b. $n \; NH_4I = n \; NH_3 = \dfrac{(0.464 \; atm)(2.0 \; L)}{(0.0821 \; L \cdot atm/mol \cdot K)(673 \; K)} = 0.0168$ mol

2.43 g NH_4I reacted

41. $(2x)^2 / (0.863 - x) = 0.144$; $x = 0.159$

$P \; NO_2 = 0.318$ atm; $P \; N_2O_4 = 0.704$ atm

43. a. (1) reverse   (2) reverse   (3) no effect
   (4) forward   (5) reverse

b. increasing the temperature will decrease K

45. a. $\leftarrow$ b. $\leftarrow$ c. no effect

47. c (could not increase by more than 0.75 atm)

49. a. $K = 0.22 \times 0.63/0.31 = 0.45$

b. $0.45 = \dfrac{(0.22 + x)(0.63 + x)}{(0.50 - x)}$; $x = 0.06$

$P \; CO = 0.28$ atm; $P \; H_2 = 0.69$ atm; $P \; H_2O = 0.44$ atm

51. $\ln K_2/K_1 = \dfrac{9400(200)}{8.31(800)(1000)}$ $K_2 = 1.33\,K_1 = 1.33(0.016) = 0.021$

53. $0.693 = \dfrac{\Delta H°(10)}{8.31(298)(308)}$ $\Delta H° = 53$ kJ

55. Will look much like present graph, starting at 1.60 atm

57.

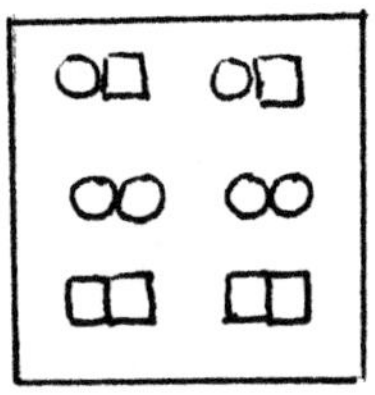

$$K = \dfrac{2^2}{2 \times 2} = 1$$

59. $P_{orig} = \dfrac{(20.0 \times 0.785\ g)(0.0821\ L \cdot atm/mol \cdot K)(453\ K)}{(60.09\ g/mol)(5.00\ L)} = 1.94$ atm

$x^2/(1.94 - x) = 0.45;$ $x = 0.74$

% left $= \dfrac{1.20}{1.94} \times 100 = 61.9$

61. a. K upside down   b. multiplied pressure by coefficient in equation

c. added instead of multiplying; used coefficients as multipliers

d. divided by volume of container

63. equation for reaction and temperature

65. $\dfrac{(x)(3x)^3}{(1.00 - 2x)^2} = 2.5$ Solving by approximations, $x \approx 0.33$

$P\ N_2 = 0.33$ atm $P\ H_2 = 0.99$ atm $P\ NH_3 = 0.34$ atm

66. $aA(g) + bB(g) \rightleftharpoons cC(g) + dD(g)$

$P\ A = [A]RT;$ $(P\ A)^a = [A]^a(RT)^a$

$K = \dfrac{(P\ C)^c \times (P\ D)^d}{(P\ A)^a \times (P\ B)^b} = \dfrac{[C]^c \times [D]^d}{[A]^a \times [B]^b} \times (RT)^{(c+d-a-b)}$

$K = K_c(RT)^{\Delta n_g}$

67. $n\ I_2$ = 0.0370 L x 0.200 mol/L x $\frac{1}{2}$ = 0.00370 mol

$n\ H_2$ = 0.00370 mol

$n\ HI$ = $\frac{3.20}{127.9}$ mol - 0.00740 mol = 0.0176 mol

$$K = \frac{P\ I_2 \times P\ H_2}{(P\ HI)^2} = \frac{n\ I_2 \times n\ H_2}{(n\ HI)^2} \times \frac{(RT/V)^2}{(RT/V)^2} = \frac{n\ I_2 \times n\ H_2}{(n\ HI)^2}$$

$$K = \frac{(0.00370)(0.00370)}{(0.0176)^2} = 0.0442$$

68. $0.45 = \frac{(x/2)^{\frac{1}{2}}(x)}{(1.00 - x)}$ By trial and error, x = 0.48

$P\ SO_3$ = 0.52 atm

69. At equilibrium:

$P\ XeF_4$ = 0.10 atm, $P\ Xe$ = 0.10 atm, $P\ F_2$ = 0.20 atm

$$K = \frac{0.10}{(0.10)(0.20)^2} = 25$$

new equilibrium pressures: $P\ XeF_4$ = 0.15 atm, $P\ Xe$ = 0.05 atm

$25 = \frac{0.15}{(0.05)(P\ F_2)^2}$ $P\ F_2$ = 0.35 atm

P initial = 0.35 atm + 2(0.15 atm) = 0.65 atm

70. Let $P_o\ I_2$ = 1.00 atm

P_{tot} = 1.40 = 1.00 - x + 2x;   x = 0.40 atm

At equilibrium: $P\ I_2$ = 0.60 atm, $P\ I$ = 0.80 atm

$K = (0.80)^2/0.60 = 1.1$

71. a. $P_o\ C_6H_5CH_2OH = \frac{(1.50/108.13\ mol)(0.0821\ L \cdot atm/mol \cdot K)(523\ K)}{2.0\ L}$

                = 0.30 atm

$x^2/(0.30 - x) = 0.56$;   x = 0.22   p = 0.22 atm

b. P_{eq} benzyl alcohol = 0.08 atm

$$\text{mass} = \frac{(108.13 \text{ g/mol})(0.08 \text{ atm})(2.0 \text{ L})}{(0.0821 \text{ L·atm/mol·K})(523 \text{ K})} = 0.4 \text{ g}$$

Saunders College Publishing

CHAPTER 13
Acids and Bases

LECTURE NOTES

This is a chapter that should be covered slowly and carefully. Students have trouble with it because there are so many different concepts. Curiously, they do better with equilibrium calculations than with qualitative concepts such as

- predicting whether a given species (in particular a salt) will give an acidic, basic, or neutral solution.

- writing a chemical equation to show why a species is acidic or basic.

To master these skills, students must know which acids and bases are strong (Chapter 4).

You will need to spend at least $2\frac{1}{2}$ lectures, probably 3, on this chapter.

LECTURE 1

I Bronsted-Lowry Model of Acids and Bases

Acid is proton donor, base is proton acceptor

$$HB(aq) + A^-(aq) \rightleftharpoons HA(aq) + B^-(aq)$$

HB, HA are acids; A^-, B^- are bases. HB - B^- and HA - A^- are conjugate acid-base pairs.

II Acidic and Basic Water Solutions

  A. In any aqueous solution, there is an equilibrium between H_3O^+ (H^+) ions and OH^- ions:

$$H_2O \rightleftharpoons H^+(aq) + OH^-(aq)$$

130

$$K_w = [H^+] \times [OH^-] = 1.0 \times 10^{-14} \text{ at } 25°C$$

1. Pure water: $[H^+] = [OH^-] = 1.0 \times 10^{-7}$ M; neutral solution

2. Acidic solution: $[H^+] > 1.0 \times 10^{-7}$ M $> [OH^-]$

3. Basic solution: $[OH^-] > 1.0 \times 10^{-7}$ M $> [H^+]$

In seawater, $[H^+] = 5 \times 10^{-9}$ M; $[OH^-] = ?$

$$[OH^-] = (1.0 \times 10^{-14})/(5 \times 10^{-9}) = 2 \times 10^{-6} \text{ M}$$

B. $pH = -\log_{10}[H^+]$

    Neutral solution: pH = 7.0
    Acidic solution: pH < 7.0
    Basic solution: pH > 7.0

    Suppose $[H^+] = 2.4 \times 10^{-6}$ M; calculate pH

$$pH = -\log_{10}(2.4 \times 10^{-6}) = 5.62$$

    Suppose pH = 8.68; calculate $[H^+]$

$$[H^+] = 10^{-8.68} = 2.1 \times 10^{-9} \text{ M}$$

    $pOH = -\log_{10}[OH^-]$; pH + pOH = 14.00 at 25°C

III <u>Weak Acids</u>

A. <u>Molecules</u>

$$HF(aq) + H_2O \rightleftharpoons H_3O^+(aq) + F^-(aq)$$

B. <u>Cations</u>

$$NH_4^+(aq) + H_2O \rightleftharpoons H_3O^+(aq) + NH_3(aq)$$

$$Zn(H_2O)_4^{2+}(aq) + H_2O \rightleftharpoons H_3O^+(aq) + Zn(H_2O)_3(OH)^+(aq)$$

<u>LECTURE 2</u>

I <u>Weak Acids</u>

A. <u>Ionization constants</u>

$$HB(aq) \rightleftharpoons H^+(aq) + B^-(aq) \qquad K_a = \frac{[H^+] \times [B^-]}{[HB]}$$

_1. Calculation of K_a

pH of 0.100 M $HC_2H_3O_2$ solution is 2.87; K_a = ?

$[H^+] = [C_2H_3O_2^-] = 1.3 \times 10^{-3}$ M

$[HC_2H_3O_2] = 0.100$ M $- 0.0013$ M $= 0.099$ M

$K_a = \dfrac{(1.3 \times 10^{-3})^2}{0.099} = 1.7 \times 10^{-5}$

2. Calculation of $[H^+]$ in solution

a. Find $[H^+]$ in 0.200 M $HC_2H_3O_2$, $K_a = 1.8 \times 10^{-5}$

Let $[H^+] = x$, then $[C_2H_3O_2^-] = x$, $[HC_2H_3O_2] = 0.200 - x$

$$\dfrac{x^2}{0.200 - x} = 1.8 \times 10^{-5}$$

To solve, assume $0.200 - x = 0.200$, in which case:

$$x^2 = 0.200(1.8 \times 10^{-5}) = 3.6 \times 10^{-6}$$

$$x = (3.6 \times 10^{-6})^{\frac{1}{2}} = 1.9 \times 10^{-3} \text{ M}$$

Approximation is generally O.K. If x < 5% of original concent-ration, O. K.

In this case, % ionization $= \dfrac{1.9 \times 10^{-3}}{0.200} \times 100 = 1.0\%$

b. Find $[H^+]$ in 0.100 M HF ($K_a = 6.9 \times 10^{-4}$)

$$\dfrac{x^2}{0.100 - x} = 6.9 \times 10^{-4}$$

Set $0.100 - x = 0.100$, solve: $x = 8.3 \times 10^{-3} > 5\%$

Make second approximation:

$$\dfrac{x^2}{0.100 - 0.008} = 6.9 \times 10^{-4}; \quad x = 8.0 \times 10^{-3}$$

II <u>Weak Bases</u>

A. <u>Molecules</u>

$$NH_3(aq) + H_2O \rightleftharpoons OH^-(aq) + NH_4^+(aq)$$

$\underline{B}$. Anions derived from weak acids

$$F^-(aq) + H_2O \rightleftharpoons HF(aq) + OH^-(aq)$$

LECTURE 3

I <u>Weak Bases</u>

A. <u>Ionization constant</u>

1. Expression for K_b:

$$NH_3: \quad K_b = \frac{[NH_4^+] \times [OH^-]}{[NH_3]}$$

2. Relation between K_a and K_b: $\quad K_a \times K_b = K_w = 1.0 \times 10^{-14}$

	K_a		K_b
HF	6.9×10^{-4}	F^-	1.4×10^{-11}
HAc	1.8×10^{-5}	Ac^-	5.6×10^{-10}
NH_4^+	5.6×10^{-10}	NH_3	1.8×10^{-5}

Strength of base is inversely related to that of conjugate weak acid.

3. Calculation of $[OH^-]$ in solution of weak base; pH of 0.10 M NaF solution?

$$\frac{[HF][OH^-]}{[F^-]} = 1.4 \times 10^{-11}; \quad \frac{[OH^-]^2}{0.10} = 1.4 \times 10^{-11}$$

$$[OH^-] = 1.2 \times 10^{-6} \text{ M}; \text{ pH} = 8.08$$

II <u>Acid-Base Properties of Salt Solutions</u>

A. <u>Cations</u>

1. Spectator: derived from strong bases

$$Li^+, \ Na^+, \ K^+; \ Ca^{2+}, \ Sr^{2+}, \ Ba^{2+}$$

2. Acidic: all other cations, including those of the transition metals.

B. <u>Anions</u>

133

1. Spectator: derived from strong acids

$$Cl^-, \ Br^-, \ I^-, \ NO_3^-, \ ClO_4^-, \ SO_4^{2-}$$

2. Basic: anions derived from weak acids, such as F^-, NO_2^-

C. <u>Overall results</u>

Salt	Cation	Anion	
$NaNO_3$	$Na^+(S)$	$NO_3^-(S)$	neutral
KF	$K^+(S)$	$F^-(B)$	basic
$FeCl_2$	$Fe^{2+}(A)$	$Cl^-(S)$	acidic

If cation is acidic, anion basic, compare K_a, K_b values

NH_4F: $K_a = 5.6 \times 10^{-10}$, $K_b = 1.4 \times 10^{-11}$; acidic

DEMONSTRATIONS

1. pH of household chemicals: SHAK <u>3</u> 65

2. Acidic properties of carbon dioxide in water: SHAK <u>3</u> 114

3. Fountain effects with ammonia, hydrogen chloride: GILB F 19; SHAK <u>3</u> 92; J. Chem. Educ. <u>72</u> 828, 931 (1995)

4. Natural indicators (grape juice, cabbage juice, tea): GILB K 23; SHAK <u>3</u> 50; J. Chem. Educ. <u>69</u> 66 (1992), <u>70</u> 326 (1993)

5. Indicator colors: GILB K 40; SHAK <u>3</u> 33, 41

6. Acid-base properties of salts: GILB K 33, M 166; SHAK <u>3</u> 103

VIDEO

1. Fountain effects: SHAK 1

2. Ammonia fountain: JCES 8

3. Cabbage juice as an indicator: SHAK 4

PROBLEMS

134

1. a. BA: H_3O^+, HCN BB: CN^-, H_2O pairs: H_3O^+, H_2O and HCN, CN^-

   b. BA: HNO_2, H_2O BB: OH^-, NO_2^- pairs: HNO_2, NO_2^- and H_2O, OH^-

   c. BA: $HC_2H_3O_2$, H_3O^+ BB: H_2O, $C_2H_3O_2^-$

      pairs: $HC_2H_3O_2$, $C_2H_3O_2^-$ and H_3O^+, H_2O

3. a. BB   b. BA   c. BA

5. a. $HC_2H_3O_2$ b. $Fe(H_2O)_6^{3+}$ c. HSO_3^-

   d. $(CH_3)_3NH^+$ e. $H_2PO_4^-$

7. a. 1.0 (A)   b. 5.00 (A)   c. -0.43 (A)   d. 8.32 (B)

9. a. $[H^+] = 1 \times 10^{-3}$ M; $[OH^-] = 1 \times 10^{-11}$ M

   b. $[H^+] = 5.9 \times 10^{-7}$ M; $[OH^-] = 1.7 \times 10^{-8}$ M

   c. $[H^+] = 5.5$ M: $[OH^-] = 1.8 \times 10^{-15}$ M

   d. $[H^+] = 2.5 \times 10^{-15}$ M; $[OH^-] = 4.0$ M

11. $[OH^-]$ in solution A = 8.7×10^{-6} M

    A is more basic; B has the higher pOH

13. $[H^+] = 1.9 \times 10^{-7}$ M; $[H^+] = 3.8 \times 10^{-7}$ M; pH = 6.42; pH = 7.02

15. a. $[H^+] = 3 \times 10^{-2}$ M

    b. $[H^+] = 8 \times 10^{-3}$ M; 30/8 = 4

17. a. $[H^+] = \dfrac{0.200/127.9}{0.400} = 3.91 \times 10^{-3}$ M; pH = 2.408

    b. $[H^+] = 1.25 \times 10^{-2}$ M; pH = 1.903; increases pH by one unit

19. n HBr = 3.50/80.91; n HNO_3 = 0.639 $\times$ 0.435

    $[H^+] = \dfrac{3.50/80.91 + 0.639 \times 0.435}{0.435} = 0.738$ M

    pH = 0.132

21. a. $[OH^-] = 0.76$ M; $[H^+] = 1.3 \times 10^{-14}$ M; pH = 13.88

    b. $[OH^-] = \dfrac{12.4/40.00}{0.750} = 0.413$ M; $[H^+] = 2.4 \times 10^{-14}$ M; pH = 13.62

23. $[OH]^- = \dfrac{2.00/74.10 + 37.5/56.11}{0.500} = 1.39$ M; $[H^+] = 7.2 \times 10^{-15}$ M

pH = 14.14

25. a. $Ni(H_2O)_3(OH)^+(aq) + H_2O \rightleftharpoons Ni(H_2O)_2(OH)_2(aq) + H_3O^+(aq)$

b. $HSO_4^-(aq) + H_2O \rightleftharpoons SO_4^{2-}(aq) + H_3O^+(aq)$

c. $HCN(aq) + H_2O \rightleftharpoons CN^-(aq) + H_3O^+(aq)$

d. $Fe(H_2O)_6^{3+}(aq) + H_2O \rightleftharpoons Fe(H_2O)_5(OH)^{2+}(aq) + H_3O^+(aq)$

e. $HC_2H_3O_2(aq) + H_2O \rightleftharpoons C_2H_3O_2^-(aq) + H_3O^+(aq)$

27. a. $HF(aq) \rightleftharpoons H^+(aq) + F^-(aq)$ $K_a = \dfrac{[H^+] \times [F^-]}{[HF]}$

b. $HSO_3^-(aq) \rightleftharpoons H^+(aq) + SO_3^{2-}(aq)$ $K_a = \dfrac{[H^+] \times [SO_3^{2-}]}{[HSO_3^-]}$

c. $HC_2H_3O_2(aq) \rightleftharpoons H^+(aq) + C_2H_3O_2^-(aq)$ $K_a = \dfrac{[H^+] \times [C_2H_3O_2^-]}{[HC_2H_3O_2]}$

29. a. 4.74   b. 9.25   c. 3.72

31. a. B < D < C < A   b. A

33. $K_a = \dfrac{(2.00 \times 10^{-2})^2}{0.230} = 1.74 \times 10^{-3}$

35. $K_a = \dfrac{(0.15)^2}{17/229} \times 0.250 = 7.6 \times 10^{-2}$

37. a. $[H^+] = (1.25 \times 2.2 \times 10^{-5})^{\frac{1}{2}} = 5.2 \times 10^{-3}$ M

b. $[H^+] = (0.66 \times 2.2 \times 10^{-5})^{\frac{1}{2}} = 3.8 \times 10^{-3}$ M

39. a. $[H^+] = (1.7 \times 6.5 \times 10^{-5})^{\frac{1}{2}} = 1.1 \times 10^{-2}$ M

b. $[OH^-] = (1.0 \times 10^{-14})/(1.1 \times 10^{-2}) = 9.1 \times 10^{-13}$ M

c. pH = 1.96

d. % ion. $= \dfrac{1.1 \times 10^{-2}}{1.7} \times 100 = 0.65\%$

Saunders College Publishing

41. $[H^+] \approx (0.60)^{\frac{1}{2}} = 0.77$ M

$\quad$ $[H^+] \approx (0.20 \times 2.2)^{\frac{1}{2}} = 0.66$ M; pH = 0.18

43. $[H^+] = (5.6 \times 10^{-10} \times 0.20)^{\frac{1}{2}} = 1.1 \times 10^{-5}$ M; pH = 4.98

45. a. $(CH_3)_2NH(aq) + H_2O \rightleftharpoons (CH_3)_2NH_2^+(aq) + OH^-(aq)$

$\quad$ b. $NO_2^-(aq) + H_2O \rightleftharpoons HNO_2(aq) + OH^-(aq)$

$\quad$ c. $H_2PO_4^-(aq) + H_2O \rightleftharpoons H_3PO_4(aq) + OH^-(aq)$

$\quad$ d. $CO_3^{2-}(aq) + H_2O \rightleftharpoons HCO_3^-(aq) + OH^-(aq)$

$\quad$ e. $HS^-(aq) + H_2O \rightleftharpoons H_2S(aq) + OH^-(aq)$

$\quad$ f. $HCO_3^-(aq) + H_2O \rightleftharpoons H_2CO_3(aq) + OH^-(aq)$

47. c < b < a < d

49. a, c

51. a. $K_a = 1 \times 10^{-8}$ $\quad$ b. $K_a = 2.6 \times 10^{-5}$

53. a. $C_3H_3O_2^-(aq) + H_2O \rightleftharpoons HC_3H_3O_2(aq) + OH^-(aq)$

$\quad$ b. 1.8×10^{-10}

$\quad$ c. $[OH^-]^2 = \dfrac{(1.8 \times 10^{-10}) \times 0.500}{94.0 \times 0.150}$; $[OH^-] = 2.5 \times 10^{-6}$ M

$\quad\quad$ $[H^+] = 4.0 \times 10^{-9}$ M; pH = 8.40

55. $[OH^-] = 4.3 \times 10^{-5}$ M; $K_b = \dfrac{(4.3 \times 10^{-5})^2}{0.002} = 9 \times 10^{-7}$

57. a. A $\quad$ b. N $\quad$ c. A $\quad$ d. B $\quad$ e. B

59. a. $Al(H_2O)_6^{3+}(aq) + H_2O \rightleftharpoons Al(H_2O)_5(OH)^{2+}(aq) + H_3O^+(aq)$

$\quad$ c. $NH_4^+(aq) + H_2O \rightleftharpoons NH_3(aq) + H_3O^+(aq)$

$\quad\quad$ $F^-(aq) + H_2O \rightleftharpoons HF(aq) + OH^-(aq)$

$\quad\quad$ Since $K_a > K_b$, 1st reaction occurs to greater extent

$\quad$ d. $CHO_2^-(aq) + H_2O \rightleftharpoons HCHO_2(aq) + OH^-(aq)$

e. $HCO_3^-(aq) + H_2O \rightleftharpoons CO_3^{2-}(aq) + H_3O^+(aq)$

$HCO_3^-(aq) + H_2O \rightleftharpoons H_2CO_3(aq) + OH^-(aq)$

2nd reaction occurs to greater extent

61. $HCl < ZnCl_2 < KCl < KF < KOH$

63. a. KF, KCN, K_2CO_3, KNO_2

b. KCl, KBr, KI, KNO_3

c. $LiClO_4$, $NaClO_4$, $KClO_4$, $Ba(ClO_4)_2$

d. NH_4ClO_4, $Fe(ClO_4)_3$, $Fe(ClO_4)_2$, $Al(ClO_4)_3$

65. (1) is weak; (2) is strong

67. 1 square, 1 circle, 9 ☐O

69. a. HCl, HF b. NH_3, CO_3^{2-}

c. K_2CO_3 d. $NaCl$

71. $[OH^-] = 1.6 \times 10^{-3}$ M; $\dfrac{(1.6 \times 10^{-3})^2}{[NH_3]} = 1.8 \times 10^{-5}$

$[NH_3] = 0.14$ M

$V = \dfrac{(0.14 \text{ mol})(0.0821 \text{ L·atm/mol·K})(298 \text{ K})}{1.00 \text{ atm}} = 3.4$ L

73. neutral

75. Measure pH of 0.10 M $AgNO_3$; if neutral, AgOH is a strong base

76. a. $H^+(aq) + OH^-(aq) \longrightarrow H_2O$; $\Delta H° = -55.8$ kJ

$\Delta H = 1.00 \text{ L} \times 0.100 \dfrac{\text{mol}}{\text{L}} \times \dfrac{-55.8 \text{ kJ}}{1 \text{ mol}} = -5.58$ kJ

b. $HF(aq) + OH^-(aq) \longrightarrow H_2O + F^-(aq)$

$\Delta H° = \Delta H_f° F^- + \Delta H_f° H_2O - \Delta H_f° OH^- - \Delta H_f° HF = -68.3$ kJ

$\Delta H = -6.83$ kJ

77. % ion. = $\dfrac{[H^+]}{[HA]_o}$; $[H^+]^2 = K_a \times [HA]_o$; $[H^+] = K_a^{\frac{1}{2}} \times [HA]_o^{\frac{1}{2}}$

$\quad$ % ion. = $K_a^{\frac{1}{2}}/[HA]_o^{\frac{1}{2}}$

78. Consider 1000 g solution

$\quad$ n $HC_2H_3O_2$ = 50.0/60.05 mol = 0.833 mol; mass water = 950 g

$\quad$ molality before ionization = 0.833 mol/0.950 kg = 0.877 m

$\quad$ V solution = 1000 g/(1.006 g/mL) = 994 mL

$\quad$ $[HC_2H_3O_2]$ = 0.833 mol/0.994 L = 0.838 M

$\quad$ $[H^+] = (0.838 \times 1.8 \times 10^{-5})^{\frac{1}{2}}$ = 3.9 x 10^{-3} M $\approx$ molality H^+

$\quad$ total molality = 0.0039 + 0.0039 + (0.877 - 0.0039) = 0.881 m

$\quad$ ΔT_f = 1.86 $\dfrac{°C}{m}$ x 0.881 m = 1.64°C $\qquad$ T_f = -1.64°C

CHAPTER 14
Equilibria in Acid-Base Solutions

LECTURE NOTES

This chapter focuses on the applications of equilibrium principles to acid-base systems. These include buffers (Section 14.1), acid-base indicators (Section 14.2), acid-base titrations (Section 14.3), and polyprotic acids (Section 14.4).

The most difficult topic is that of buffers. Students have a great deal of trouble calculating the pH of a buffer after addition of strong acid or base. Such a calculation requires that they apply two different concepts: stoichiometry and equilibrium principles.

Two lectures should be sufficient for this chapter. The first one concentrates on buffers, the second on acid-base titrations.

LECTURE 1

I <u>Buffers</u>

   A. Prepared by adding both the weak acid HB and its conjugate base B^- to water.

$$[H^+] = K_a \times \frac{[HB]}{[B^-]} = K_a \times \frac{n\ HB}{n\ B^-}$$

Calculate $[H^+]$ in a solution prepared by adding 0.200 mol of acetic acid and 0.200 mol of sodium acetate to one liter.

$$[H^+] = 1.8 \times 10^{-5} \times \frac{0.200}{0.200} = 1.8 \times 10^{-5}\ M \qquad pH = 4.74$$

   B. <u>Choice of system</u> Note that since HB and B^- must be present in roughly equal amounts, $[H^+]$ of the buffer is roughly equal to K_a of the weak acid. To establish a buffer of pH 7, choose a system such as $H_2PO_4^-$, HPO_4^{2-}, where $K_a = 6.2 \times 10^{-8}$.

To prepare a solution of pH 7.00:

At pH 7.00, $\dfrac{[H_2PO_4^-]}{[HPO_4^{2-}]} = \dfrac{1.0 \times 10^{-7}}{6.2 \times 10^{-8}} = 1.6$

Add 160 mL of 0.100 M $H_2PO_4^-$ to 100 mL of 0.100 M HPO_4^{2-}

C. <u>Effect of adding strong acid or base to buffer</u>

$$HB(aq) + OH^-(aq) \longrightarrow H_2O + B^-(aq)$$

$$B^-(aq) + H^+(aq) \longrightarrow HB(aq)$$

As a result of these reactions, H^+ or OH^- ions are consumed, so do not drastically change the pH.

Calculate pH of buffer containing 0.200 mol of HAc and 0.200 mol of Ac^- after addition of 0.020 mol NaOH.

	orig. conc.	change	final conc.
HAc	0.200	−0.020	0.180
Ac^-	0.200	+0.020	0.220

$$[H^+] = 1.8 \times 10^{-5} \times \frac{0.180}{0.220} = 1.5 \times 10^{-5}$$

pH = 4.82, as compared to 4.74 originally

II <u>Indicators</u> Derived from weak acid HIn

$$\frac{[HIn]}{[In^-]} = \frac{[H^+]}{K_a \text{ of HIn}}$$

Color depends upon the ratio of concentrations of HIn, In^-, which depends on $[H^+]$ and K_a of indicator. For bromthymol blue, $K_a = 10^{-7}$, HIn is yellow, In^- is blue.

pH < 6; $[HIn] > 10[In^-]$; solution is yellow

pH > 8; $[HIn] < 0.1[In^-]$; solution is blue

pH = 7; $[HIn] = [In^-]$; solution is green

<u>LECTURE 2</u>

I <u>Acid-Base titrations</u>

   A. <u>Strong acid - strong base</u>

$$H^+(aq) + OH^-(aq) \longrightarrow H_2O \qquad K = 1/K_w = 1.0 \times 10^{14}$$

     pH = 7 at end point, changes rapidly near end point

   B. <u>Weak acid - strong base</u>

$$HA(aq) + OH^-(aq) \longrightarrow H_2O + A^-(aq) \qquad K = 1/K_b$$

     If $K_a = 1 \times 10^{-5}$, $K_b = 1 \times 10^{-9}$, $K = 1 \times 10^9$

     Start with 0.10 M HA: $[H^+] = (1 \times 10^{-6})^{\frac{1}{2}} = 1 \times 10^{-3}$; pH = 3.0

     Half-way to end point: $[HA] = [A^-]$; $[H^+] = 1 \times 10^{-5}$; pH = 5.0

     At end point $\approx$ 0.10 M B^-; $[OH^-]^2 = 1 \times 10^{-10}$; pH = 9.0

     In general, pH > 7 at end point, changes slowly near end point

   C. <u>Strong acid - weak base</u>

$$H^+(aq) + A^-(aq) \longrightarrow HA(aq) \qquad K = 1/(K_a \text{ of HA})$$

     pH < 7 at end point, changes slowly near end point.

   D. <u>Choice of indicator</u>
      1. Weak acid - strong base Use indicator such as PP, which changes color above pH 7

      2. Strong acid - weak base Use indicator such as MR, which changes color below pH 7

      3. Strong acid - strong base Any indicator OK because pH changes so rapidly at end point.

II <u>Polyprotic Acids</u>

   Virtually all of H^+ comes from 1st dissociation

$$H_2CO_3(aq) \rightleftharpoons H^+(aq) + HCO_3^-(aq) \qquad K_1 = 4 \times 10^{-7}$$

$$HCO_3^-(aq) \rightleftharpoons H^+(aq) + CO_3^{2-}(aq) \qquad K_2 = 4 \times 10^{-11}$$

   In 0.10 M H_2CO_3, $[H^+] = [HCO_3^-] = 2 \times 10^{-4}$ M, $[CO_3^{2-}] = 4 \times 10^{-11}$ M

142

DEMONSTRATIONS

1. Buffer action: GILB K 30, K 32, K 47; SHAK $\underline{3}$ 173; J. Chem. Educ. $\underline{72}$ 345 (1995)
2. Buffer capacity: GILB K 48
3. HCO_3^- - CO_3^{2-} buffer: GILB K 41
4. Acetic acid - acetate buffer: GILB K 50
5. Buffering action of alka-seltzer: SHAK $\underline{3}$ 186
6. Acid-base neutralization: GILB K 40: SHAK $\underline{1}$ 15
7. Titration curves: SHAK $\underline{3}$ 167
8. pH change during acid-base titration: GILB K 37, K 39
9. Carbon dioxide and limewater: SHAK $\underline{1}$ 329

VIDEO

1. Strong acids, weak acids, and buffers: JCES 28
2. Limewater and carbon dioxide: Falcon 29

PROBLEMS

1. a. $H^+(aq) + OH^-(aq) \longrightarrow H_2O$

   b. $H^+(aq) + NH_3(aq) \longrightarrow NH_4^+(aq)$

   c. $HNO_2(aq) + F^-(aq) \rightleftharpoons HF(aq) + NO_2^-(aq)$

   d. $HCN(aq) + OH^-(aq) \longrightarrow CN^-(aq) + H_2O$

3. a. $H^+(aq) + C_2H_3O_2^-(aq) \longrightarrow HC_2H_3O_2(aq)$

   b. $H^+(aq) + OH^-(aq) \longrightarrow H_2O$

   c. $H^+(aq) + SO_3^{2-}(aq) \longrightarrow HSO_3^-(aq)$

5. a. $1/(1.0 \times 10^{-14}) = 1.0 \times 10^{14}$

   b. $1/(5.6 \times 10^{-10}) = 1.8 \times 10^{9}$

   c. $K_a\ HNO_2/K_a\ HF = (6.0 \times 10^{-4})/(6.9 \times 10^{-4}) = 0.87$

   d. $1/K_b\ CN^- = 1/(1.7 \times 10^{-5}) = 5.9 \times 10^{4}$

7. a. $K = 1/(1.8 \times 10^{-5}) = 5.6 \times 10^{4}$

   b. 1.0×10^{14}

   c. $K = 1/(6.0 \times 10^{-8}) = 1.7 \times 10^{7}$

9. $[H^+] = \dfrac{1.8 \times 10^{-5}}{0.500} \times [HC_2H_3O_2]$

   a. $[H^+] = 3.6 \times 10^{-6}$ M; pH = 5.44

   b. $[H^+] = 9.0 \times 10^{-6}$ M; pH = 5.05

   c. $[H^+] = 1.8 \times 10^{-5}$ M; pH = 4.74

   d. $[H^+] = 2.7 \times 10^{-5}$ M; pH = 4.57

11. $[H^+] = 6.6 \times 10^{-5} \times \dfrac{(0.075)(0.310)}{0.250} = 6.1 \times 10^{-6}$ M

    pH = 5.21

13. a. Benzoic acid - sodium benzoate

    b. Hypochlorous acid - sodium hypochlorite

    c. Ammonium chloride - ammonia

15. a. $[H^+] = 3.2 \times 10^{-4} = (6.0 \times 10^{-4}) \times \dfrac{[HNO_2]}{[NO_2^-]}$

    ratio = 0.53

    b. $n\ HNO_2 = 0.53 \times 0.42 = 0.22$ mol

    c. $n\ NaNO_2 = 0.750 \times 0.250/0.53$

    mass $NaNO_2 = \dfrac{0.750 \times 0.250}{0.53} \times 69.0$ g = 24 g

17. a. $[H^+] = \dfrac{(6.9 \times 10^{-4})(5.00/20.01)}{(10.0/41.99)} = 7.2 \times 10^{-4}$ M; pH = 3.14

b. 3.14

19. $5.6 \times 10^{-5} = \dfrac{(1.8 \times 10^{-5})(0.417)}{n\ KC_2H_3O_2}$

$n\ KC_2H_3O_2 = \dfrac{1.8(0.417)}{5.6}$; mass $= \dfrac{(1.8)(0.417)}{5.6} \times 98.9$ g = 13 g

21. $K_a = \dfrac{(5.1 \times 10^{-5})^2}{0.171} = 1.5 \times 10^{-8}$

$[H^+] = 1.5 \times 10^{-8} \times \dfrac{0.171}{0.200} = 1.3 \times 10^{-8}$ M; pH = 7.89

23. a. $[H^+] = 4.7 \times 10^{-11} \times \dfrac{0.15}{0.20} = 3.5 \times 10^{-11}$ M; pH = 10.46

b. $[H^+] = 4.7 \times 10^{-11} \times \dfrac{0.18}{0.17} = 5.0 \times 10^{-11}$ M; pH = 10.30

c. $[H^+] = 4.7 \times 10^{-11} \times \dfrac{0.12}{0.23} = 2.5 \times 10^{-11}$ M; pH = 10.60

25. same; dilution should not change pH of buffer

27. a. $[H^+] = 9.4 \times 10^{-11}$ M; pH = 10.03

b. $\dfrac{[HCO_3^-]}{[CO_3^{2-}]} = [H^+]/K_a = (3.2 \times 10^{-10})/(4.7 \times 10^{-11}) = 6.8$

29. c. Contains 0.20 mol HNO_2, 0.60 mol NO_2^-

None of the others contains a weak acid and its conjugate weak base.

31. n butyric acid $= \dfrac{2.00\ g}{88.1\ g/mol} = 0.0227$; n NaOH $= \dfrac{0.50\ g}{40.0\ g/mol} = 0.0125$

0.0102 mol butyric acid, 0.0125 mol butyrate ion

$[H^+] = 1.55 \times 10^{-5} \times \dfrac{0.0102}{0.0125} = 1.26 \times 10^{-5}$ M

pH = 4.898

33. n lactic acid = $\dfrac{10.00 \text{ g}}{90.08 \text{ g/mol}}$ = 0.1110 mol

n KOH = $\dfrac{8.00 \text{ g}}{56.11 \text{ g/mol}}$ = 0.143 mol

0.032 mol KOH in excess

$[OH^-] = \dfrac{0.032 \text{ mol}}{0.475 \text{ L}}$ = 0.067 M; pOH = 1.17; pH = 12.83

35. a. $[H_2PO_4^-]/[HPO_4^{2-}] = [H^+]/K_a = (4.0 \times 10^{-8})/(6.2 \times 10^{-8}) = 0.65$

   b. ratio = $(1.0 \times 10^{-7})/(6.2 \times 10^{-8})$ = 1.6

      start with 0.65 mol $H_2PO_4^-$, 1.00 mol HPO_4^{2-}

      end with 1.02 mol $H_2PO_4^-$, 0.63 mol HPO_4^{2-}

           about 37%

   c. ratio = $(3.2 \times 10^{-8})/(6.2 \times 10^{-8})$ = 0.52

      end with 0.56 mol $H_2PO_4^-$, 1.09 mol HPO_4^{2-}

           about 14%

37. a. PP   b. MO   c. MO   d. any

39. pK_a = 3.45; range is approximately 3.5 - 4.5; orange

41. a. 25.00 mL x $\dfrac{0.3000}{0.5125}$ = 14.63 mL

   b. pOH = 0.52; pH = 13.48

   c. $[OH^-] = \dfrac{(0.02500)(0.3000)}{2(0.0323)}$ = 0.116 M; pOH = 0.94; pH = 13.06

   d. pH = 7.00

43. a. $H^+(aq) + C_2H_3O_2^-(aq) \longrightarrow HC_2H_3O_2(aq)$

   b. before: Na^+, $C_2H_3O_2^-$

      halfway: Na^+, $C_2H_3O_2^-$, Cl^-, $HC_2H_3O_2$

      equivalence: Na^+, Cl^-, $HC_2H_3O_2$

   c. 25.00 mL x 0.2000/0.1500 = 33.33 mL

d. before: $[OH^-]^2 = (5.6 \times 10^{-10})(0.20)$; $[OH^-] = 1.06 \times 10^{-5}$ M

pH = 9.02

halfway: $[H^+] = K_a \, HC_2H_3O_2 = 1.8 \times 10^{-5}$ M; pH = 4.74

equivalence: $[HC_2H_3O_2] = \dfrac{(0.02500)(0.2000)}{0.05833} = 0.0857$ M

$[H^+]^2 = (1.8 \times 10^{-5})(0.0857)$; $[H^+] = 1.2 \times 10^{-3}$ M

pH = 2.91

45. a. $[F^-] = \dfrac{5.00/42.00}{1.00} = 0.119$ M; $K_b = 1.4 \times 10^{-11}$

$[OH^-]^2 = (1.4 \times 10^{-11})(0.119) = 1.3 \times 10^{-6}$ M; pOH = 5.89

pH = 8.11

b. $[H^+] = K_a \, HF = 6.9 \times 10^{-4}$ M; pH = 3.16

c. n HF = $\dfrac{5.00 \times 25.00}{42,000} = 2.98 \times 10^{-3}$

V HBr = $\dfrac{0.119 \times 25.00}{0.250} = 11.9$ mL

$[HF] = \dfrac{2.98 \times 10^{-3} \text{ mol}}{0.0369 \text{ L}} = 0.0807$ M

$[H^+]^2 = (6.9 \times 10^{-4})(0.0807)$; $[H^+] = 0.0075$ M; pH = 2.13

47. $H_3PO_4(aq) \rightleftharpoons 3H^+(aq) + PO_4^{3-}(aq)$

$K = K_1K_2K_3 = (7.1 \times 10^{-3})(6.2 \times 10^{-8})(4.5 \times 10^{-10})$

$= 2.0 \times 10^{-22}$

49. $[H^+]^2 \approx (4.4 \times 10^{-7})(0.25)$; $[H^+] = 3.3 \times 10^{-4}$ M; pH = 3.48

$[HCO_3^-] = [H^+] = 3.3 \times 10^{-4}$ M

$[CO_3^{2-}] = K_a \, HCO_3^- = 4.7 \times 10^{-11}$ M

51. $[H_2CO_3] = 0.04$ M, $[HCO_3^-] = 0.16$ M, $[CO_3^{2-}] \approx 0$

53.

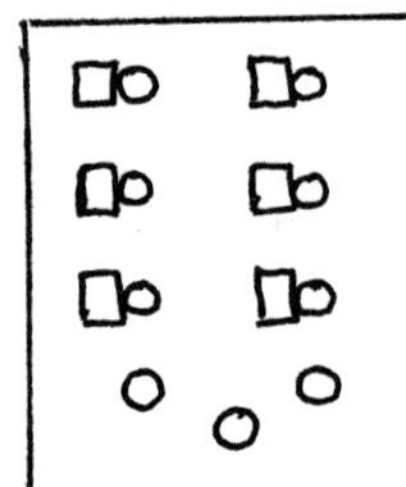   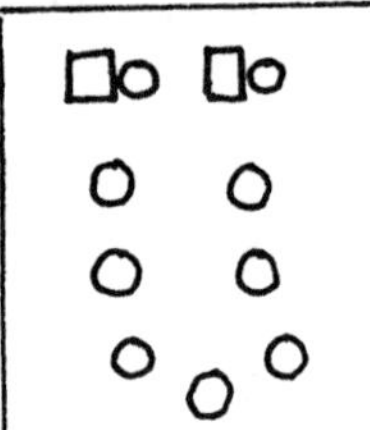   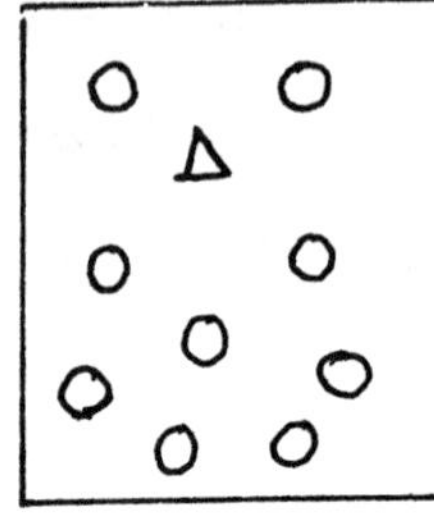

   a. buffer   b. buffer   c. base

55. a. weak   b. $\approx 10^{-4}$ c. 9

57. a. $[H^+] = K_a \times [HLac]/[Lac^-]$;   adding HLac increases $[H^+]$

b. NH_3 is a weak base

c. reacts with both H^+ and OH^-

d. could be weak acid; "low" is ambiguous

59. excess $OH^- = 2(0.03800(0.125) - (0.04500)(0.1250) = 0.00388$ mol

$$[OH^-] = \frac{3.88 \times 10^{-3} \text{ mol}}{0.0830 \text{ L}} = 0.0467 \text{ M}; \quad pOH = 1.33; \quad pH = 12.67$$

61. n $Ac^- = $ n $OH^- = (0.0750)(1.50) = 0.112$ mol

n HAc = n $Ac^- \times \dfrac{[H^+]}{K_a} = (1.0 \times 10^{-4})(0.112)/(1.8 \times 10^{-5}) = 0.62$

Add $(0.62 + 0.11)$ mol HAc $= 0.73$ mol HAc

$$\frac{(V \times 1.0542)0.98}{60.0} = 0.73; \quad 42 \text{ mL}$$

62. $K_a = 5.6 \times 10^{-10}$; $pK_a = 9.25$

Buffer would contain very low concentration of ammonia; would have almost no capacity for absorbing acid.

63. $pK_a = 4.26$; $K_a = 5.5 \times 10^{-5}$; $[H^+]_{orig.} = 2.8 \times 10^{-3}$ M

$$\frac{(2.8 \times 10^{-3})^2}{4.0/0.250 \, \mathcal{M}} = 5.5 \times 10^{-5}$$

Solving, $\mathcal{M} = 1.1 \times 10^2$ g/mol

64. a. $[H^+]^2 = 1.8 \times 10^{-5}$; $[H^+] = 4.2 \times 10^{-3}$ M; pH = 2.37

b. $[H^+] = 1.8 \times 10^{-5}$ M; pH = 4.74

c. $[H^+] = 1.8 \times 10^{-5} \times \dfrac{0.10}{49.90} = 3.6 \times 10^{-8}$ M; pH = 7.44

d. 0.500 M $NaC_2H_3O_2$

$[OH^-]^2 = 5.6 \times 10^{-10} \times 0.50$; $[OH^-] = 1.7 \times 10^{-5}$ M; pH = 9.22

e. $[OH^-] = 1.000$ M $\times \dfrac{0.10}{100.1} = 1.0 \times 10^{-3}$ M; pH = 11.00

f. $[OH^-] = 1.000$ M $\times \dfrac{50.00}{150.00} = 0.333$ M; pH = 13.52

65. $1.0 \times 10^{-5} = 1.8 \times 10^{-5} \times \dfrac{[HC_2H_3O_2]}{[C_2H_3O_2^-]}$

$\dfrac{[HC_2H_3O_2]}{[C_2H_3O_2^-]} = 0.56$; fraction neutralized $= \dfrac{1.00}{1.56} = 0.64$

32 mL NaOH

66. $\log_{10}[H^+] = \log_{10}K_a = \log_{10}[HB]/[B^-]$

multiply by -1: pH $= pK_a + \log_{10}[B^-]/[HB]$

67. a. $[H^+] = 0.1500$ M; pH = 0.82

b. $1.1 \times 10^{-2} = \dfrac{[H^+] \times [SO_4^{2-}]}{[HSO_4^-]} = \dfrac{(0.1500 + x)(x)}{0.1500 - x}$

Solving by successive approximations, $x \approx 0.01$

$[H^+] = 0.16$ M; pH = 0.80

CHAPTER 15
Complex Ions

LECTURE NOTES

One thing to keep in mind throughout this chapter is the amount of "jargon" involved. Such terms as "chelating agent", "coordination number", "square planar" and "ligand" are unfamiliar to students. They also have trouble visualizing the structure of a complex given only its formula, e.g., $Co(en)_2(NH_3)_2^{3+}$.

This is a short chapter with two main topics:

- geometry of complexes, including isomerism. Here it helps to emphasize the geometric relationships in an octahedron, a figure which is unfamiliar to many students.

- electronic structure of complexes. Here we discuss only the crystal field model and restrict that to octahedral complexes. Before doing that, we review briefly the electronic structure of transition metal ions, covered way back in Chapter 6.

Two lectures should be quite adequate for this chapter; this could be cut to 1½ lectures if you're running behind.

<u>LECTURE 1</u>

I <u>Complex Ions and Coordination Compounds</u>

    A. A complex ion consists of a central metal cation (usually derived from a transition metal) joined by coordinate covalent bonds to two or more molecules or anions called ligands.

Complex ion	Cation	Ligands	Coord. No.
$Ag(NH_3)_2^+$	Ag^+	2 NH_3 molecules	2
$Cu(H_2O)_4^{2+}$	Cu^{2+}	4 H_2O molecules	4
$Fe(CN)_6^{3-}$	Fe^{3+}	6 CN^- ions	6

Reaction between cation and ligand can be considered to be an acid-base reaction:

$$Ag^+(aq) \; + \; 2NH_3(aq) \; \longrightarrow \; Ag(NH_3)_2^+(aq)$$

Lewis acid   Lewis base

(Lewis acid accepts electron pair, Lewis base donates it)

Coordination compound contains complex ion. Examples:

$$[Cu(H_2O)_4]SO_4 \qquad [Ag(NH_3)_2]NO_3 \qquad K_3[Fe(CN)_6]$$

Charge of central cation in $Zn(H_2O)_3(OH)^+$?

$$+1 = -1 + X; \; X = +2$$

B. Nature of ligands: ordinarily contain at least one unshared pair of electrons.

$$H - \overset{..}{N} - H \qquad H - \overset{..}{\underset{..}{O}} - H \qquad (:\overset{..}{\underset{..}{O}} - H)^- \qquad (:\overset{..}{\underset{..}{Cl}}:)^-$$
$$| \\ H$$

If the ligand contains two or more unshared pairs on different, non-adjacent atoms, it can act as a chelating agent, forming more than one bond with the central metal ion.

$$H - \overset{..}{N} - CH_2 - CH_2 - \overset{..}{N} - H \qquad \text{ethylenediamine (en)}$$
$$| \qquad\qquad\qquad | \\ H \qquad\qquad\qquad H$$

Forms complex ions such as $Cu(en)_2^{2+}$, $Cr(en)_3^{3+}$

II <u>Geometry of Complex Ions</u>

  A. <u>Coordination no. = 2</u>: linear, $180°$ bond angle

$$(NH_3 - Ag - NH_3)^+$$

  B. <u>Coordination no. = 4</u>

    1. Tetrahedral $Zn(NH_3)_4^{2+}$

    2. Square planar $Cu(NH_3)_4^{2+}$

  c. <u>Coordination no. = 6</u>

    Octahedral: $Co(NH_3)_6^{3+}$

D. <u>Geometric isomerism</u>

1. Square planar

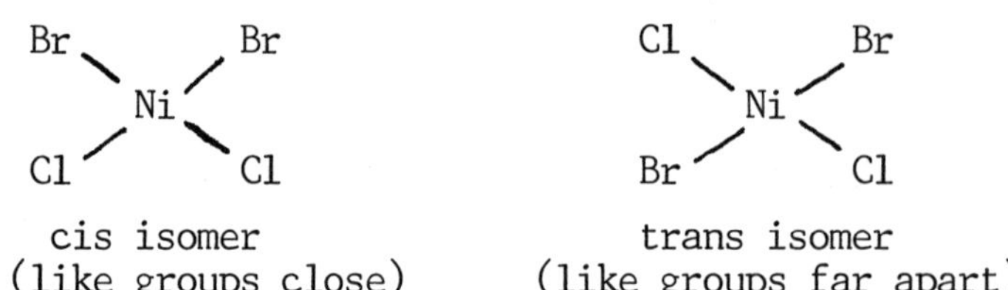

cis isomer   trans isomer
(like groups close)   (like groups far apart)

B. Octahedral: $Co(NH_3)_4Cl_2^+$

trans   cis

<u>LECTURE 2</u>

I <u>Electronic Structure Transition Metal Ions</u>

No outer s electrons

Cr^{3+} [Ar] $3d^3$ Co^{2+} [Ar] $3d^7$ Zn^{2+} [Ar] $3d^{10}$

II <u>Crystal Field Model</u>

Approach of six ligands to transition metal cation splits d orbitals into two sets of different energy:

high energy: $d_{x^2-y^2}$, d_{z^2}

low energy: d_{xy}, d_{yz}, d_{xz}

A. Splitting enery Δ_o, small: Hund's rule is followed, giving "high-spin complex" with the maximum number of unpaired electrons.

Co^{2+} (↑)(↑)   3 unpaired electrons
            (↑↓)(↑↓)(↑)

B. Splitting energy large: electrons fill lower-energy orbitals, giving "low-spin complex".

Co^{2+} $(\uparrow\,)(\quad)$ 1 unpaired electron

$(\uparrow\downarrow)(\uparrow\downarrow)(\uparrow\downarrow)$

  C. Color   Caused by electron transitions between two sets of d orbitals

III <u>Formation Constants</u>

$$Ag^{+}(aq) + 2NH_3(aq) \rightleftharpoons Ag(NH_3)_2^{+}(aq) \qquad K_f = 1.7 \times 10^{7}$$

$$Ag^{+}(aq) + 2S_2O_3^{2-}(aq) \rightleftharpoons Ag(S_2O_3)_2^{3-}(aq) \qquad K_f = 1 \times 10^{13}$$

$S_2O_3^{2-}$ complex more stable than that with NH_3

DEMONSTRATIONS

1. Paramagnetism of complexes: GILB L 34

2. Complexing agents: GILB S 3

3. Complexes (and precipitates) of Ag^{+}: SHAK <u>1</u> 307; J. Chem. Educ. <u>57</u> 813 (1980)

4. Complexes of cobalt(II): GILB E 10, J 35; SHAK <u>1</u> 280; J. Chem. Educ. <u>57</u> 453 (1980)

5. Complexes (and precipitates) of Cu^{2+}: GILB E 9, J 38; SHAK <u>1</u> 314, 318

6. Complexes (and precipitates) of Fe^{3+}: GILB J 2; SHAK <u>1</u> 338

7. Complexes of Hg^{2+} (orange tornado): SHAK <u>1</u> 271

8. Complexes (and precipitates) of Ni^{2+}: GILB J 30; SHAK <u>1</u> 299; J. Chem. Educ. <u>57</u> 900 (1980)

9. Complexes (and precipitates) of Pb^{2+}: SHAK <u>1</u> 286

<u>VIDEO</u>

1. Complexes and precipitates of Ag^{+}: SHAK 13

2. Iodo complexes of Hg^{2+}: SHAK 12

3. Complexes and precipitates of Ni^{2+}: SHAK 14

PROBLEMS

1. a. $C_2O_4^{2-}$, NH_3, Cl^- b. +2   c. $Na_3[Ni(C_2O_4)_2(NH_3)Cl]$

3. a. $[Fe(CO)_4Cl_2]^+$ b. $[Fe(en)_2]^{3+}$ c. $[Fe(SCN)_2(H_2O)_2(OH)_2]^-$

5. a. 6   b. 4   c. 4   d. 2

7. a. $[Ag(CN)_2]^-$ b. $[Pt(H_2O)_4]^{2+}$ c. $[Cd(CN)_4]^{2-}$ d. $[Fe(C_2O_4)_3]^{3-}$

9. $\mathcal{M} = (107.9 + 64.14 + 24.02 + 28.02)$ g/mol = 224.1 g/mol

$$\frac{64.14}{224.1} \times 100\% = 28.62\%$$

11. $4(55.85$ g/mol$) = 0.0035(\mathcal{M})$; $\mathcal{M} = 6.9 \times 10^4$ g/mol

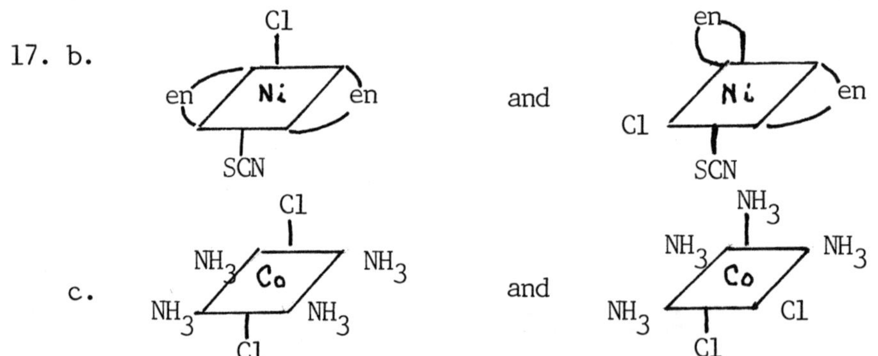

13.

15.

17. b.

c.

Saunders College Publishing

19.

$$
\begin{array}{ccc}
\text{(Fe with Cl, NH}_3\text{, Br, NH}_3\text{, NH}_3\text{, Cl)} &
\text{(Fe with Br, NH}_3\text{, Cl, NH}_3\text{, NH}_3\text{, Cl)} &
\text{(Fe with NH}_3\text{, NH}_3\text{, Cl, Br, NH}_3\text{, Cl)}
\end{array}
$$

21. a. $1s^2 2s^2 2p^6 3s^2 3p^6 4s^2 3d^{10} 4p^6 4d^8$

    b. $1s^2 2s^2 2p^6 3s^2 3p^6 3d^3$

    c. $1s^2 2s^2 2p^6 3s^2 3p^6 4s^2 3d^{10} 4p^6 5s^2 4d^{10} 5p^6 4f^{14} 5d^6$

    d. $1s^2 2s^2 2p^6 3s^2 3p^6 3d^3$

    e. $1s^2 2s^2 2p^6 3s^2 3p^6 4s^2 3d^{10} 4p^6 4d^4$

23.

    4d

    a. [Kr] (↑↓)(↑↓)(↑↓)(↑)(↑)   2 unp. e⁻

    3d

    b. [Ar] (↑)(↑)(↑)( )( )   3 unp. e⁻

                    4f   5d

    c. [Xe] (↑↓)(↑↓)(↑↓)(↑↓)(↑↓)(↑↓)(↑↓) (↑↓)(↑)(↑)(↑)(↑) 4 unp. e⁻

    3d

    d. [Ar] (↑)(↑)(↑)( )( )   3 unp. e⁻

    e. [Kr] (↑)(↑)(↑)(↑)( )   4 unp. e⁻

25.

           (↑↓)(↑↓)   (↑)(↑)   ( )( )

    a. (↑↓)(↑↓)(↑↓)   b. (↑↓)(↑)(↑) and (↑↓)(↑↓)(↑↓)

27. 4 d electrons in Mo^{2+}, only 3 in Mo^{3+}

        (↑)( )   ( )( )

      (↑)(↑)(↑) ←— Mo^{2+} —→ (↑↓)(↑)(↑)

29.

           ( )( )   (↑)(↑)

    Co(CN)$_6^{3-}$: (↑↓)(↑↓)(↑↓)   Co(H$_2$O)$_6^{3+}$: (↑↓)(↑)(↑)

31. a. 4 b. 4   c. 4   d. 2   e. 4 (Hund's rule followed with all)

155

33. $E = \dfrac{(6.626 \times 10^{-34} \text{ J·s})(2.998 \times 10^{8} \text{ m/s})}{(399 \times 10^{-9} \text{ m})} \times \dfrac{1 \text{ kJ}}{10^{3} \text{ J}} \times \dfrac{6.022 \times 10^{23}}{1 \text{ mol}}$

$= 3.00 \times 10^{2}$ kJ/mol

35. Absorbs in green at about 500 nm

37. $Zn^{2+}(aq) + 4\ OH^{-}(aq) \rightleftharpoons Zn(OH)_4^{2-}(aq)$

$K_f = 3 \times 10^{14} = \dfrac{[Zn(OH)_4^{2-}]}{[Zn^{2+}][OH^{-}]^4}$

$\dfrac{[Zn^{2+}]}{[Zn(OH)_4^{2-}]} = \dfrac{3.3 \times 10^{-15}}{[OH^{-}]^4}$

pH	1.0	7.0	10.0
pOH	13.0	7.0	4.0
$[OH^{-}]^4$	1×10^{-52}	1×10^{-28}	1×10^{-16}
ratio	3×10^{37}	3×10^{13}	30

39. a. $[Ag^{+}]/[Ag(NH_3)_2^{+}] = 1/(1.7 \times 10^{7})[NH_3]^2 = 1$; $[NH_3] = 2.4 \times 10^{-4}$ M

b. $[Ni^{2+}]/[Ni(NH_3)_6^{2+}] = 1/(9 \times 10^{8})[NH_3]^6 = 1$; $[NH_3] = 3 \times 10^{-2}$ M

41. a. Forms stable complex with Fe^{3+}

b. All positions are equivalent

c. They accept electron pairs from ligands

43. a. False; 6, not 5

b. False; 0 unpaired electrons

c. False; shorter wavelength

45. a. n Pt = 0.270; n N = 0.54; n H = 1.62; n Cl = 1.078

$PtN_2H_6Cl_4$

b. [diagram of two octahedral Pt complexes: left with NH$_3$, Cl, Cl, Cl, Cl, NH$_3$ ligands; right with NH$_3$, Cl, NH$_3$, Cl, Cl, Cl ligands]

47. $2.0 \times 10^2 = \dfrac{[\text{Hem CO}]}{[\text{Hem O}_2]} \times P\,O_2/P\,CO$

$P\,CO/P\,O_2 = \dfrac{[\text{Hem CO}]}{[\text{Hem O}_2]} \times 1/200 = \dfrac{0.050}{0.950} \times 1/200 = 2.6 \times 10^{-4}$

49. $\mathcal{M}\ \text{Na}_4(\text{EDTA}) = [4(22.99) + 10(12.01) + 2(14.01) + 8(16.00) + 12(1.008)]$

$= 380.2 \text{ g/mol}$

$n\ \text{Na}_4(\text{EDTA}) = n\ \text{Pb} = 0.50/207.2$

$\text{mass} = \dfrac{0.50}{207.2} \times 380.2 \text{ g} = 0.92 \text{ g}$

50. $\text{Pt(NH}_3)_4^{2+}$ cation, PtCl_4^{2-} anion <u>or</u>

$\text{Pt(NH}_3)_3\text{Cl}^+$ cation, $\text{Pt(NH}_3)\text{Cl}_3^-$ anion

51. a. $n\ \text{Cu} = 0.3186$; $n\ \text{C} = 1.273$; $n\ \text{H} = 7.01$; $n\ \text{N} = 1.917$; $n\ \text{S} = 0.3190$

$n\ \text{O} = 1.274$

simplest formula : $\text{CuC}_4\text{H}_{22}\text{N}_6\text{SO}_4$

b. noting that ethylenediamine has the molecular formula $\text{C}_2\text{H}_8\text{N}_2$, the cation must be $\text{Cu(en)}_2(\text{NH}_3)_2^{2+}$; the complex ions are cis and trans isomers

52. $55\ \dfrac{\text{kcal}}{\text{mol}} \times \dfrac{4.18\text{kJ}}{1\text{ kcal}} \times \dfrac{10^3\ \text{J}}{1\text{ kJ}} \times \dfrac{1\text{ mol}}{6.022 \times 10^{23}} = 3.82 \times 10^{-19}\ \text{J}$

$\lambda = \dfrac{(6.626 \times 10^{-34}\ \text{J·s})(2.998 \times 10^8\ \text{m/s})}{3.82 \times 10^{-19}\ \text{J}} = 5.2 \times 10^{-7}\ \text{m} = 520\ \text{nm}$

absorbs in green; color should be red-violet

CHAPTER 16
Precipitation Equilibria

LECTURE NOTES

This is a relatively short chapter which can be covered in 1 - 1½ lectures, depending on whether you discuss

-the calculation of solubilities in strong acid or a complexing agent (Examples 16.7, 16.9)

-qualitative analysis (Section 16.3)

In general, students find K_{sp} calculations straightforward, perhaps easier than for any other type of equilibrium constant. Some of them, however, confuse K_{sp} with s, the molar solubility.

LECTURE 1

I. <u>Solubility Equilibrium</u>; K_{sp}

  A. Expression for K_{sp}

$$AgCl(s) \rightleftharpoons Ag^+(aq) + Cl^-(aq)$$

$$K_{sp} = [Ag^+] \times [Cl^-]$$

$$PbCl_2(s) \rightleftharpoons Pb^{2+}(aq) + 2Cl^-(aq)$$

$$K_{sp} = [Pb^{2+}] \times [Cl^-]^2 = 1.7 \times 10^{-5}$$

  B. Uses of K_{sp}

    1. Calculation of concentration of one ion, knowing that of the other.

    What is $[Pb^{2+}]$ in a solution in equilibrium with lead chloride if $[Cl^-] = 0.020$ M?

$$[Pb^{2+}] = (1.7 \times 10^{-5})/(2.0 \times 10^{-2})^2 = 0.042 \text{ M}$$

2. Determination of whether precipitate will form.

If $P < K_{sp}$, no precipitate (equilibrium not established)

If $P > K_{sp}$, precipitate forms until P becomes equal to K_{sp}

Suppose enough Ag^+ is added to a solution 0.001 M in CrO_4^{2-} to make $[Ag^+]$ = 0.001 M. Will silver chromate precipitate (K_{sp} $Ag_2CrO_4 = 2 \times 10^{-12}$)?

$$P = (\text{orig. conc. } Ag^+)^2 \times (\text{orig. conc. } CrO_4^{2-})$$

$$= (1 \times 10^{-3})^2 \times (1 \times 10^{-3}) = 1 \times 10^{-9} > K_{sp}$$

precipitate forms

3. Water solubility

$$PbCl_2(s) \rightleftharpoons Pb^{2+}(aq) + 2Cl^-(aq)$$

$$[Pb^{2+}] = s; \quad [Cl^-] = 2s$$

$$4s^3 = K_{sp} = 1.7 \times 10^{-5}; \quad s = 1.6 \times 10^{-2} \text{ M}$$

Solubility in grams per 100 mL?

$$1.6 \times 10^{-2} \frac{\text{mol}}{\text{L}} \times \frac{278.1 \text{ g}}{1 \text{ mol}} \times \frac{1 \text{ L}}{10^3 \text{ mL}} \times 10^2 \text{ mL} = 0.44 \text{ g}$$

II <u>Dissolving Precipitates</u>

A. <u>Strong acid</u> (works with basic anions)

$$CaCO_3(s) + 2H^+(aq) \longrightarrow Ca^{2+}(aq) + H_2CO_3(aq)$$

B. NH_3 or OH^- (works with cation that forms complexes)

$$AgCl(s) + 2NH_3(aq) \longrightarrow Ag(NH_3)_2^+(aq) + Cl^-(aq)$$

$$K = K_{sp} \times K_f = (1.8 \times 10^{-10})(1.7 \times 10^7) = 3.1 \times 10^{-3}$$

Solubility of AgCl in 1.00 M NH_3?

$$\frac{s^2}{1.00} = 3.1 \times 10^{-3}; \quad s = 0.056 \text{ mol/L}$$

DEMONSTRATIONS

1. Determination of K_{sp}: GILB J 27
2. Will a precipitate form? Will it dissolve? J. Chem. Educ. 71 69 (1994)
3. Formation and dissolving of precipitates: GILB J 34
4. Fractional precipitation: GILB J 23; J. Chem. Educ. 65 359 (1988)
5. Common ion effect: GILB J 4; J. Chem. Educ. 70 155 (1993)
6. Equilibrium between AgCl and Ag_2CrO_4: J. Chem. Educ. 54 618 (1977)
7. Solubility of silver salts: GILB J 24
8. Reaction of Hg^{2+} with I^-: GILB I 15
9. Amphoterism: GILB M 65; SHAK 3 128

(see also # 3, 5, 6, 8, Chapter 15)

VIDEO

1. Precipitation of HgI_2: JCES 29

PROBLEMS

1. a. $Ba(C_2H_3O_2)_2(s) \rightleftharpoons Ba^{2+}(aq) + 2C_2H_3O_2^-(aq)$

$$K_{sp} = [Ba^{2+}] \times [C_2H_3O_2^-]^2$$

b. $Al_2(SO_4)_3(s) \rightleftharpoons 2Al^{3+}(aq) + 3SO_4^{2-}(aq)$

$$K_{sp} = [Al^{3+}]^2 \times [SO_4^{2-}]^3$$

c. $Zr(OH)_4(s) \rightleftharpoons Zr^{4+}(aq) + 4\ OH^-(aq)$

$$K_{sp} = [Zr^{4+}] \times [OH^-]^4$$

d. $Mn(SO_4)_2(s) \rightleftharpoons Mn^{4+}(aq) + 2SO_4^{2-}(aq)$

$$K_{sp} = [Mn^{4+}] \times [SO_4^{2-}]^2$$

3. a. $TiO_2(s) \rightleftharpoons Ti^{4+}(aq) + 2\,O^{2-}(aq)$

   b. $K_2Cr_2O_7(s) \rightleftharpoons 2K^+(aq) + Cr_2O_7^{2-}(aq)$

   c. $Al_2S_3(s) \rightleftharpoons 2Al^{3+}(aq) + 3S^{2-}(aq)$

   d. $PbC_2O_4(s) \rightleftharpoons Pb^{2+}(aq) + C_2O_4^{2-}(aq)$

5. $K_{sp}\ Mg_3(PO_4)_2 = 1 \times 10^{-24}$

   a. $[Mg^{2+}]^3 = (1 \times 10^{-24})/(3 \times 10^{-5})^2 = 1 \times 10^{-15}$

   $[Mg^{2+}] = 1 \times 10^{-5}\ M$

   b. $[PO_4^{3-}]^2 = (1 \times 10^{-24})/(2 \times 10^{-3})^3 = 1 \times 10^{-16}$

   $[PO_4^{3-}] = 1 \times 10^{-8}\ M$

   c. $8[PO_4^{3-}]^5 = 1 \times 10^{-24};\ [PO_4^{3-}] = 1 \times 10^{-5}\ M$

   d. $9[Mg^{2+}]^5 = 1 \times 10^{-24};\ [Mg^{2+}] = 1 \times 10^{-5}\ M$

7. a. $[Mg^{2+}] = (6 \times 10^{-12})/(2.0 \times 10^{-4})^2 = 2 \times 10^{-4}\ M$

   b. $[Fe^{2+}] = (5 \times 10^{-17})/(2.0 \times 10^{-4})^2 = 1 \times 10^{-9}\ M$

   c. $[Fe^{3+}] = (3 \times 10^{-39})/(2.0 \times 10^{-4})^3 = 4 \times 10^{-28}\ M$

9. a. $[Ba^{2+}] = \dfrac{1.8 \times 10^{-7}}{(0.025)^2} = 2.9 \times 10^{-4}\ M$

   b. $[F^-]^2 = (1.8 \times 10^{-7})/0.200 = 9.0 \times 10^{-7};\ [F^-] = 9.5 \times 10^{-4}\ M$

   % left $= \dfrac{9.5 \times 10^{-4}}{0.025} \times 100 = 3.8\%;\ 96.2\%$ precipitated

11. $(1 \times 10^{-3})(5 \times 10^{-4}) = 5 \times 10^{-7} > 1.8 \times 10^{-8};$ yes

   $[Pb^{2+}] = (1.8 \times 10^{-8})/(1 \times 10^{-3}) = 2 \times 10^{-5}\ M$

13. a. $[Fe^{2+}] = \dfrac{25.0}{70.0} \times 0.075\ M = 0.027\ M$

   pOH $= 5.50$; orig. $[OH^-] = 3.2 \times 10^{-6}\ M$

   $[OH^-] = 3.2 \times 10^{-6}\ M \times \dfrac{45.0}{70.0} = 2.1 \times 10^{-6}\ M$

   $(2.7 \times 10^{-2})(2.1 \times 10^{-6})^2 = 1.2 \times 10^{-13} > K_{sp};$ yes

b. $[Fe^{2+}] \approx 0.027$ M, since it is in large excess

$$[OH^-]^2 = \frac{5 \times 10^{-17}}{0.027} \; ; \quad [OH^-] = 4 \times 10^{-8} \text{ M}$$

$$[Na^+] = 2.1 \times 10^{-6} \text{ M}; \quad [NO_3^-] = 2[Fe^{2+}] = 0.054 \text{ M}$$

15. solubility $Ba(IO_3)_2 = \dfrac{27 \times 10^{-3} \text{ g}}{487 \text{ g/mol} \times 0.100 \text{ L}} = 5.5 \times 10^{-4}$ mol/L

$$K_{sp} = 4s^3 = 4(5.5 \times 10^{-4})^3 = 6.7 \times 10^{-10}$$

17. a. $Ca_3(PO_4)_2(s) \rightleftharpoons 3Ca^{2+}(aq) + 2PO_4^{3-}(aq)$

$$108s^5 = 1 \times 10^{-33} \; ; \quad s = 1 \times 10^{-7} \text{ M}$$

$$1 \times 10^{-7} \frac{\text{mol}}{\text{L}} \times \frac{310 \text{ g}}{1 \text{ mol}} = 3 \times 10^{-5} \text{ g/L}$$

b. $[PO_4^{3-}]^2 = \dfrac{1 \times 10^{-33}}{1 \times 10^{-9}} \; ; \quad [PO_4^{3-}] = 1 \times 10^{-12}$ mol/L

$$s = 5 \times 10^{-13} \text{ mol/L}$$

$$5 \times 10^{-13} \frac{\text{mol}}{\text{L}} \times \frac{310 \text{ g}}{1 \text{ mol}} = 2 \times 10^{-10} \text{ g/L}$$

c. $[Ca^{2+}]^3 = \dfrac{1 \times 10^{-33}}{4 \times 10^{-4}} \; ; \quad [Ca^{2+}] = 1.4 \times 10^{-10}$ M

$$s = [Ca^{2+}]/3 = 4.5 \times 10^{-11} \text{ M}$$

$$4.5 \times 10^{-11} \frac{\text{mol}}{\text{L}} \times \frac{310 \text{ g}}{1 \text{ mol}} = 1 \times 10^{-8} \text{ g/L}$$

19. $[Pb^{2+}] = \dfrac{1.00/278}{1.0} = 3.6 \times 10^{-3}$ M

$$[Cl^-] = 7.2 \times 10^{-3} \text{ M}$$

$$P = (3.6 \times 10^{-3})(7.2 \times 10^{-3})^2 = 1.9 \times 10^{-7} < K_{sp} \; ; \quad \text{no}$$

21. a. $FeS(s) + 2H^+(aq) \longrightarrow Fe^{2+}(aq) + H_2S(aq)$

b. $Bi(OH)_3(s) + 3H^+(aq) \longrightarrow Bi^{3+}(aq) + 3H_2O$

c. $Cu(NH_3)_4^{2+}(aq) + 4H^+(aq) \longrightarrow Cu^{2+}(aq) + 4NH_4^+(aq)$

d. $MgCO_3(s) + 2H^+(aq) \longrightarrow Mg^{2+}(aq) + H_2CO_3(aq)$

23. a. $AgCl(s) + 2NH_3(aq) \longrightarrow Ag(NH_3)_2^+(aq) + Cl^-(aq)$

   b. $Al^{3+}(aq) + 3NH_3(aq) + 3H_2O \longrightarrow Al(OH)_3(s) + 3NH_4^+(aq)$

   c. $Cu^{2+}(aq) + 4NH_3(aq) \longrightarrow Cu(NH_3)_4^{2+}(aq)$

25. a. $Sb^{3+}(aq) + 3\ OH^-(aq) \longrightarrow Sb(OH)_3(s)$

   b. $Sb(OH)_3(s) + OH^-(aq) \longrightarrow Sb(OH)_4^-(aq)$

   c. $Sb^{3+}(aq) + 4\ OH^-(aq) \longrightarrow Sb(OH)_4^-(aq)$

27. $Al(OH)_3(s) + 3H^+(aq) \rightleftharpoons Al^{3+}(aq) + 3H_2O$

$$K = (K_{sp}\ Al(OH)_3)/(K\ H_2O)^3 = (2 \times 10^{-31})/(1 \times 10^{-14})^3$$
$$= 2 \times 10^{11}$$

29. $Al(OH)_3(s) \rightleftharpoons Al^{3+}(aq) + 3\ OH^-(aq)$

$Al^{3+}(aq) + 4\ OH^-(aq) \rightleftharpoons Al(OH)_4^-(aq)$

$Al(OH)_3(s) + OH^-(aq) \rightleftharpoons Al(OH)_4^-(aq)$

$$K = K_{sp} \times K_f = 2 \times 10^2$$

31. $[OH^-] = 1 \times 10^{-3}$

$$2 \times 10^2 = [Al(OH)_4^-]/(1 \times 10^{-3})\ ; \quad s = [Al(OH)_4^-] = 0.2\ M$$

33. $AgI(s) + 2NH_3(aq) \rightleftharpoons Ag(NH_3)_2^+(aq) + I^-(aq)$

$$K = K_{sp}\ AgI \times K_f\ Ag(NH_3)_2^+ = (1 \times 10^{-16})(1.7 \times 10^7) = 2 \times 10^{-9}$$
$$2 \times 10^{-9} = \frac{s^2}{(3.0)^2}\ ; \quad s = 1 \times 10^{-4}\ M$$

35. a. $K = K_{sp}\ Zn(OH)_2 \times K_f\ Zn(OH)_4^{2-}$

$$= (4 \times 10^{-17})(3 \times 10^{14}) = 1 \times 10^{-2}$$

   b. $s = \dfrac{1.00/99.4}{1.00} = 0.0101\ mol/L$

$$1 \times 10^{-2} = \frac{1 \times 10^{-2}}{[OH^-]^2}\ ; \quad [OH^-] = 1\ M$$

37. a. I, HCl, AgCl

b. II, H_2S, H^+, Bi_2S_3

c. III, H_2S, OH^-, CoS

d. IV, $(NH_4)_2CO_3$, $MgCO_3$

39. Ag^+, Pb^{2+}, Hg_2^{2+}; Ba^{2+}, Ca^{2+}, Mg^{2+}

41. CdS

43. 7 squares, 7 circles distributed evenly through box

45. 2 MX_2 solid, 4 M cations in solution, 8 X anions in solution

47. $[F^-] = \dfrac{1.0/19.0 \text{ mol}}{1000 \text{ L}} = 5.3 \times 10^{-5}$ M

$P = (2.0 \times 10^{-4})(5.3 \times 10^{-5})^2 = 5.6 \times 10^{-13} < K_{sp}$; no

49. $Mg(OH)_2(s) \rightleftharpoons Mg^{2+}(aq) + 2\,OH^-(aq)$; $K_1 = K_{sp}\ Mg(OH)_2$

$2NH_4^+(aq) \rightleftharpoons 2H^+(aq) + 2NH_3(aq)$; $K_2 = (K_a\ NH_4^+)^2$

$2H^+(aq) + 2\,OH^-(aq) \rightleftharpoons 2H_2O$ $K_3 = 1/K_w^2$

$K = K_1 K_2 K_3 = \dfrac{(6 \times 10^{-12})(5.6 \times 10^{-10})^2}{(1.0 \times 10^{-14})^2} = 2 \times 10^{-2}$

$4s^3 = (0.02)(0.20)^2$; $s = 0.06$ M

water: $4s^3 = 6 \times 10^{-12}$; $s = 1 \times 10^{-4}$ M

50. $K = K_{sp}\ CaF_2/(K_a\ HF)^2 = \dfrac{1.5 \times 10^{-10}}{(6.9 \times 10^{-4})^2} = 3.2 \times 10^{-4}$

$[H^+] = 1.9 \times 10^{-4}$ M $\times\ 0.30/0.20 = 2.8 \times 10^{-4}$ M

$4s^3/(2.8 \times 10^{-4})^2 = 3.2 \times 10^{-4}$; $s = 1.8 \times 10^{-4}$ M

51. $[Ag^+] = (1.8 \times 10^{-10})/0.020 = 9.0 \times 10^{-9}$ M

$[I^-] = (1 \times 10^{-16})/(9 \times 10^{-9}) = 1 \times 10^{-8}$ M

52. a. $[OH^-]^2 = \dfrac{6 \times 10^{-12}}{5.6 \times 10^{-2}} = 1 \times 10^{-10}$ $[OH^-] = 1 \times 10^{-5}$ M

   b. $Al(OH)_3$: $P = (4 \times 10^{-7})(1 \times 10^{-5})^3 = 4 \times 10^{-22} > K_{sp}$

      $Fe(OH)_3$: $P = (2 \times 10^{-7})(1 \times 10^{-5})^3 = 2 \times 10^{-22} > K_{sp}$

      $Al(OH)_3$ and $Fe(OH)_3$ precipitate

   c. $[OH^-]^2 = \dfrac{6 \times 10^{-12}}{2.8 \times 10^{-2}}$ $[OH^-] = 1 \times 10^{-5}$ M

      $[Al^{3+}] = \dfrac{2 \times 10^{-31}}{1 \times 10^{-15}} = 2 \times 10^{-16}$ M; $[Fe^{3+}] = 3 \times 10^{-24}$ M

      virtually all

   d. 0.028 mol $Mg(OH)_2 + 4 \times 10^{-7}$ mol $Al(OH)_3 + 2 \times 10^{-7}$ mol $Fe(OH)_3$

      0.028 mol $\times \dfrac{58.32 \text{ g}}{1 \text{ mol}} = 1.6$ g

53. a. $K = (K_f\ Zn(OH)_4^{2-})/(K_f\ Zn(NH_3)_4^{2+}) = (3 \times 10^{14})/(3.6 \times 10^8)$

      $= 8 \times 10^5$

   b. $[OH^-]^2 = 1.8 \times 10^{-5}$ $[OH^-]^4 = 3.2 \times 10^{-10}$

      $\dfrac{[Zn(OH)_4^{2-}]}{[Zn(NH_3)_4^{2+}]} = \dfrac{8 \times 10^5 \times 3.2 \times 10^{-10}}{(1.0)^4} = 2.6 \times 10^{-4}$

      $\dfrac{[Zn(NH_3)_4^{2+}]}{[Zn(OH)_4^{2-}]} = 4 \times 10^3$

54. $Ag(NH_3)_2^+(aq) \rightleftharpoons Ag^+(aq) + 2NH_3(aq)$ $K = 1/K_f$

     $2NH_3(aq) + 2H^+(aq) \rightleftharpoons 2NH_4^+(aq)$ $K = 1/K_a^2$

     $Ag^+(aq) + Cl^-(aq) \rightleftharpoons AgCl(s)$ $K = 1/K_{sp}$

     $K = \dfrac{1}{K_f K_a^2 K_{sp}} = 1.0 \times 10^{21}$

CHAPTER 17
Spontaneity of Reaction

LECTURE NOTES

This chapter takes an empirical approach to the thermodynamic functions H, S and G. Emphasis is placed on the effect of $\Delta H°$, $\Delta S°$, and $\Delta G°$ on reaction spontaneity. We do not attempt to derive the Gibbs-Helmholtz equation, nor do we discuss how molar entropy is determined. These aspects of thermodynamics can safely be deferred to a later course. It is important that students, in their first exposure, appreciate the power of thermodynamics, even at the expense of its rigor or elegance.

Perhaps surprisingly, students find this material quite straight-forward. It is readily covered in two lectures.

LECTURE 1

I Spontaneous Reactions

    A. Examples:

$$CH_4(g) + 2\ O_2(g) \longrightarrow CO_2(g) + 2H_2O(l)$$

$$H_2O(s) \longrightarrow H_2O(l) \qquad (at\ 25°C)$$

    B. Factors affecting

        1. Energy factor: at 25°C, 1 atm, exothermic reactions are ordinarily spontaneous ($\Delta H < 0$).

        2. Randomness factor: other things being equal, system tends to move from a more ordered to a more random state.

II Entropy Changes

    A. $\Delta S = S$ products − S reactants; measure of change in order

        solid $\longrightarrow$ liquid; ΔS positive

        liquid $\longrightarrow$ gas : ΔS positive

166

Saunders College Publishing

ΔS is usually positive for a reaction in which the number of moles of gas increases.

$$2SO_3(g) \longrightarrow 2SO_2(g) + O_2(g); \quad \Delta n_g = +1; \quad \Delta S \text{ positive}$$

$$N_2(g) + 3H_2(g) \longrightarrow 2NH_3(g) ; \quad \Delta n_g = -2; \quad \Delta S \text{ negative}$$

B. Calculation of $\Delta S°$ (ΔS at 1 atm, 1 M)

$$\Delta S° = \Sigma S° \text{ products} - \Sigma S° \text{ reactants}$$

$$Fe_2O_3(s) + 3H_2(g) \longrightarrow 2Fe(s) + 3H_2O(g)$$

$$\Delta S° = 2S° \, Fe(s) + 3S° \, H_2O(g) - S° \, Fe_2O_3(s) - 3S° \, H_2(g)$$

$$= 2(27.2 \text{ J/K}) + 3(188.7 \text{ J/K}) - 90.0 \text{ J/K} - 3(130.6 \text{ J/K})$$

$$= 138.7 \text{ J/K}$$

Note that $S°$ is a positive quantity for both compounds and elements.

C. Reactions for which $\Delta S°$ is positive tend to be spontaneous, at least at high temperatures

$$H_2O(s) \longrightarrow H_2O(l) \quad \Delta S° > 0$$

$$H_2O(l) \longrightarrow H_2O(g) \quad \Delta S° > 0$$

$$Fe_2O_3(s) + 3H_2(g) \longrightarrow 2Fe(s) + 3H_2O(g) \quad \Delta S° > 0$$

All of these reactions are endothermic ($\Delta H > 0$). They become spontaneous at high temperatures.

III <u>Free Energy Changes</u>

A. $\Delta G° = \Delta H° - T\Delta S°$

Note that ΔG, like ΔS, is dependent on pressure, concentration. Unlike $\Delta H°$ and $\Delta S°$, $\Delta G°$ is strongly temperature dependent because of T in equation.

If $\Delta G° < 0$, reaction spontaneous at standard conditions (T, P)

If $\Delta G° > 0$, reaction nonspontaneous at standard conditions

If $\Delta G° = 0$, reaction at equilibrium at standard conditions

B. Effect of $\Delta H°$, $\Delta S°$ on spontaneity

If $\Delta H° > 0$, $\Delta S° < 0$, $\Delta G° > 0$ at all T, nonspontaneous

If $\Delta H° < 0$, $\Delta S° > 0$, $\Delta G° < 0$ at all T, spontaneous

If $\Delta H° > 0$, $\Delta S° > 0$, $\Delta G° > 0$ at low T, < 0 at high T

spontaneous at high T

If $\Delta H° < 0$, $\Delta S° < 0$, $\Delta G° < 0$ at low T, > 0 at high T

nonspontaneous at high T

LECTURE 2

I <u>Free Energy Change</u>

A. Calculation of $\Delta G°$ from $\Delta H°$ and $\Delta S°$

1. $Fe_2O_3(s) + 3H_2(g) \longrightarrow 2Fe(s) + 3H_2O(g)$

$\Delta H° = +96.8$ kJ; $\Delta S° = +138.7$ J/K $= +0.1387$ kJ/K

$\Delta G°$ (in kJ) $= +96.8 - 0.1387$ T

at 25°C: $\Delta G° = +96.8 - 0.1387(298) = +55.5$ kJ

at 500°C: $\Delta G° = +96.8 - 0.1387(773) = -10.4$ kJ

nonspontaneous at 25°C, spontaneous at 500°C

2. At what temperature does the reduction of Fe_2O_3 by hydrogen become spontaneous at 1 atm?

$\Delta G° = 0$; $\Delta H° = T\Delta S°$

$T = \Delta H°/\Delta S° = 96.8/0.1387 = 698$ K (425°C)

B. Calculation of $\Delta G°$ at 25°C from $\Delta G_f°$

$\Delta G° = \sum \Delta G_f°$ products $- \sum \Delta G_f°$ reactants

$PbCl_2(s) \longrightarrow Pb^{2+}(aq) + 2Cl^-(aq)$

$\Delta G° = \Delta G_f° \, Pb^{2+} + 2\,\Delta G_f° \, Cl^- - \Delta G_f° \, PbCl_2 = +27.3$ kJ

C. Calculation of ΔG from $\Delta G°$

$\Delta G = \Delta G° + RT \ln Q$

$PbCl_2(s) \longrightarrow Pb^{2+}(0.0010 \text{ M}) + 2Cl^-(0.0010 \text{ M})$

$\Delta G = \Delta G° + (0.00831)(298) \ln (0.0010)^3 = -24.0$ kJ

Note that reaction is spontaneous at this low concentration

D. Relation between $\Delta G°$ and K

$$\Delta G° = -RT \ln K$$

If $K > 1$, $\Delta G° < 0$, reaction spontaneous at standard conditions

If $K < 1$, $\Delta G° > 0$, reaction nonspontaneous at standard cond.

If $K = 1$, $\Delta G° = 0$, reaction at equilibrium at standard cond.

$$H_2O(l) \longrightarrow H^+(aq) + OH^-(aq)$$

$$\Delta G° = -(0.00831)(298) \ln (1.0 \times 10^{-14}) = +79.8 \text{ kJ}$$

Nonspontaneous at standard conditions ($1 M H^+$, $1 M OH^-$)

DEMONSTRATIONS

1. Thermodynamic changes: GILB H 60
2. Spontaneous endothermic reactions: GILB H 19, H 20
3. Entropy change: GILB H 61, H 64
4. Free energy change: GILB H 66
5. Maximum work: J. Chem. Educ. <u>67</u> 962 (1990)
6. Decomposition of methanol: J. Chem. Educ. <u>70</u> 585 (1993)
7. Greenhouse effect: J. Chem. Educ. <u>70</u> 773 (1993)

VIDEO

1. Reaction of CuO with hydrogen: SHAK 9

PROBLEMS

1. d

3. d

5. a. +   b. +   c. -   d. -

7. a. $\Delta n_g = +1$; +   b. $\Delta n_g = -2$; -   c. $\Delta n_g = -1$; -   d. $\Delta n_g = +3$; +

169

9. a. $\Delta n_g = +1$; +   b. $\Delta n_g = +1$; +   c. −

11. a. $\Delta S° = 26.9$ J/K + 373.6 J/K − 69.9 J/K − 89.6 J/K = +241.0 J/K

    b. $\Delta S° = 112.1$ J/K − 70.4 J/K − 213.6 J/K = −171.9 J/K

    c. $\Delta S° = 769.2$ J/K + 1435.0 J/K − 960.0 J/K − 419.4 J/K = +824.8 J/K

    d. $\Delta S° = 854.4$ J/K + 419.4 J/K − 1435.0 J/K − 459.0 J/K = −620.2 J/K

13. a. $\Delta S° = -10.8$ J/K − 69.9 J/K = −80.7 J/K

    b. $\Delta S° = 106.0$ J/K + 348.3 J/K + 279.6 J/K − 382.4 J/K − 667.8 J/K

       = −316.3 J/K

    c. $\Delta S° = 56.5$ J/K − 336.3 J/K + 209.7 J/K − 162.3 J/K − 124.8 J/K

       = −357.2 J/K

15. a. $\Delta S° = 20.1$ J/K − 69.9 J/K − 256.7 J/K = −306.5 J/K

    b. $\Delta S° = 282.5$ J/K + 90.4 J/K − 279.6 J/K − 364.5 J/K = −271.2 J/K

    c. $\Delta S° = -128.9$ J/K + 248.1 J/K + 139.8 J/K − 20.1 J/K − 29.9 J/K

       = +209.0 J/K

17. a. $\Delta G° = 136$ kJ − 318(0.0390)kJ = + 124 kJ

    b. $\Delta G° = -713$ kJ − 318(0.1180)kJ = −751 kJ

    c. $\Delta G° = 12.9$ kJ + 318(0.2161)kJ = +81.6 kJ

19. a. $\Delta H° = -601.7$ kJ − 184.6 kJ + 285.8 kJ + 641.3 kJ = +140.8 kJ

       $\Delta G° = 140.8$ kJ − 425(0.2410)kJ = +38 kJ; no

    b. $\Delta H° = -1216.3$ kJ + 553.5 kJ + 393.5 kJ = −269.3 kJ

       $\Delta G° = -269.3$ kJ + 425(0.1719)kJ = −196.2 kJ; yes

    c. $\Delta H° = -184.4$ kJ + 1714.8 kJ − 132.8 kJ = +1397.6 kJ

       $\Delta G° = 1397.6$ kJ − 425(0.8248)kJ = +1047 kJ; no

    d. $\Delta H° = -1574.0$ kJ − 1714.8 kJ + 169.4 kJ = −3119.4 kJ

       $\Delta G° = -3119.4$ kJ + 425(0.6202)kJ = −2856 kJ; yes

21. a. $\Delta G° = -157.2$ kJ $+ 237.2$ kJ $= +80.0$ kJ

b. $\Delta G° = -930.4$ kJ -948.8 kJ $+ 894.4$ kJ $+ 309.6$ kJ $= -675.2$ kJ

c. $\Delta G° = -131.2$ kJ $- 441.3$ kJ $- 711.6$ kJ $+ 8.0$ kJ $= -1276.1$ kJ

23. a. $Zn(s) + I_2(s) \longrightarrow ZnI_2(s)$

$\Delta H° = -208.0$ kJ; $\Delta S° = (0.1611 - 0.1161 - 0.0416)$ kJ/K $= 0.0034$ kJ/K

$\Delta G°_f = -208.0$ kJ $- 298(0.0034)$ kJ $= -209.0$ kJ/mol

b. $C(s) + \frac{1}{2}H_2(g) + 3/2\ Cl_2(g) \longrightarrow CHCl_3(1)$

$\Delta H° = -134.5$ kJ; $\Delta S° = (0.2017 - 0.3345 - 0.0653 - 0.0057)$ kJ/K

$= -0.2038$ kJ/K

$\Delta G°_f = -134.5$ kJ $+ 298(0.2038)$ kJ $= -73.8$ kJ/mol

c. $Ag(s) + \frac{1}{2}N_2(g) + 3/2\ O_2(g) \longrightarrow AgNO_3(s)$

$\Delta H° = -124.4$ kJ

$\Delta S° = (0.1409 - 0.3075 - 0.0958 - 0.0426)$ kJ/K $= -0.3050$ kJ/K

$\Delta G°_f = -124.4$ kJ $+ 298(0.3050)$ kJ $= -33.5$ kJ/mol

25. $CH_3OH(1) \longrightarrow CH_4(g) + \frac{1}{2}\ O_2(g)$

$\Delta G° = \Delta G°_f\ CH_4(g) - \Delta G°_f\ CH_3OH(1)$

$= 50.7$ kJ $+ 166.3$ kJ $= +115.6$ kJ;   no

27. a. $C_2H_4(g) + 2HCl(g) + \frac{1}{2}\ O_2(g) \longrightarrow C_2H_4Cl_2(1) + H_2O(1)$

b. $\Delta S° = \dfrac{\Delta H° - \Delta G°}{T} = \dfrac{-318.7\ kJ + 194.4\ kJ}{298\ K} = -0.417$ kJ/K

c. $-0.417 = S°\ C_2H_4Cl_2 + 0.0699$ kJ/K $- 0.1025$ kJ/K $- 0.3736$ kJ/K

$- 0.2195$ kJ/K

$S°\ C_2H_4Cl_2 = 0.209$ kJ/K

29. a. $\Delta S° = \dfrac{-353.2\ kJ + 452.4\ kJ}{298\ K} = +0.334$ kJ/K ; yes, Δn_g is $+$

b. $+0.334$ kJ/K $= 2S° \text{ COCl}_2 + 0.3736$ kJ/K $- 0.2050$ kJ/K $- 0.4034$ kJ/K

$S° \text{ COCl}_2 = +0.284$ kJ/K

c. -353.2 kJ $= 2\Delta H°_f \text{ COCl}_2 - 184.6$ kJ $+ 269.0$ kJ

$\Delta H°_f \text{ COCl}_2 = -218.8$ kJ

31. a. spontaneous up to about 3000 K

b. nonspontaneous at all T

c. nonspontaneous below about 1050 K

33. a. $1648.4/0.5494 \approx 3000$ K

b. never

c. $197.8/0.1878 \approx 1050$ K

35. $\Delta H° = -1130.7$ kJ $- 393.5$ kJ $- 241.8$ kJ $+ 1901.6$ kJ $= +135.6$ kJ

$\Delta S° = 135.0$ J/K $+ 213.6$ J/K $+ 188.7$ J/K $- 203.4$ J/K $= +333.9$ J/K

$T = 135.6/0.3339 = 406$ K $= 133°C$

spontaneous above 133°C

37. a. $\Delta H° = -1180.5$ kJ $+ 1648.4$ kJ $= +467.9$ kJ

$\Delta S° = (+0.1092 + 0.6408 - 0.0171 - 0.1748)$kJ/K $= 0.5581$ kJ/K

$\Delta G° = 467.9$ kJ $- 0.5581$ T

T(K)	100	200	300	400	500
$\Delta G°$(kJ)	412.1	356.3	300.5	244.7	188.8

b. $T = 467.9/0.5581 = 838$ K $= 565°C$

39. a. $SnO_2(s) \longrightarrow Sn(s) + O_2(g)$

$\Delta H° = +580.7$ kJ; $\Delta S° = +0.2043$ kJ/K; $T = 2840$ K

b. $SnO_2(s) + 2H_2(g) \longrightarrow Sn(s) + 2H_2O(g)$

$\Delta H° = +97.1$ kJ; $\Delta S° = +0.1155$ kJ/K; $T = 841$ K

c. $SnO_2(s) + C(s) \longrightarrow Sn(s) + CO_2(g)$

$\Delta H° = +187.2$ kJ; $\Delta S° = 0.2072$ kJ/K; $T = 903$ K

(b) occurs at lowest T

41. graphite $\rightleftharpoons$ diamond

$\Delta H° = +1.9$ kJ; $\Delta S° = 0.0024 - 0.0057 = -0.0033$ kJ/K

Not at equilibrium at any T at 1 atm

43. $I_2(s) \rightleftharpoons I_2(g)$

$\Delta H° = 62.4$ kJ; $\Delta S° = 260.6$ J/K $- 116.1$ J/K $= +144.5$ J/K

$T = 62.4/0.1445 = 432$ K $= 159°C$

45. a. no; $\Delta G = +27.2$ kJ

b. $\Delta G = 27.2$ kJ $+ 0.00831(298)$ ln $(3.0 \times 10^{-3})^2 = -1.6$ kJ; yes

47. a. $\Delta G° = -314.4$ kJ $+ 208.0$ kJ $+ 474.4$ kJ $= +368.0$ kJ

b. $\Delta G = 368.0$ kJ $+ 0.00831(298)$ ln $\dfrac{(5.4 \times 10^{-5})^2(0.100)}{(0.75)^2}$

$= 368.0$ kJ $- 52.9$ kJ $= +315.1$ kJ

49. a. $\Delta G° = -553.6$ kJ $- 557.6$ kJ $+ 1167.3$ kJ $= +56.1$ kJ

b. $-1.0 = +56.1 + 0.00831(298)$ ln $4s^3$

$s = 3 \times 10^{-4}$ M

$[Ca^{2+}] = 3 \times 10^{-4}$ M, $[F^-] = 6 \times 10^{-4}$ M

51. a. $\Delta G° = -32.9$ kJ $+ 101.4$ kJ $= +68.5$ kJ; nonspontaneous

b. $2CH_4(g) + \frac{1}{2} O_2(g) \longrightarrow C_2H_6(g) + H_2O(g)$

$\Delta G° = -32.9$ kJ $- 228.6$ kJ $+ 101.4$ kJ $= -160.1$ kJ

spontaneous

53. $C_6H_{12}O_6(aq) + 6 O_2(g) + 80ADP(aq) + 80HPO_4^{2-}(aq) + 160H^+(aq) \longrightarrow$

$6 CO_2(g) + 86H_2O + 80ADP(aq)$; $\Delta G° = = -390$ kJ

173

55. a. $\Delta G° = 2\Delta G_f^o$ HCl $- 2\Delta G_f^o$ HI $= -190.6$ kJ $- 3.4$ kJ $= -194.0$ kJ

    b. $\ln K = \dfrac{194.0}{(8.31 \times 10^{-3})(298)} = 78.3$; $K = 1 \times 10^{34}$

57. a. $\ln K = -(68.6)/(8.31 \times 10^{-3})(298)$; $K = 9 \times 10^{-13}$

    b. $\ln K = (46.3)/(8.31 \times 10^{-3})(1200)$; $K = 1 \times 10^{2}$

59. $MgSO_4(s) \rightleftharpoons Mg^{2+}(aq) + SO_4^{2-}(aq)$

    $\Delta G° = -454.8$ kJ $- 744.5$ kJ $+ 1170.7$ kJ $= -28.6$ kJ

    $\ln K_{sp} = (28.6)/(8.31 \times 10^{-3})(298)$; $K_{sp} = 1 \times 10^{5}$

61. a. An exothermic reaction is spontaneous at low temperatures

    b. – – – cannot occur at standard conditions.

    c. – – – usually positive – – – number of moles of gas

    d. If $\Delta H°$ is negative and $\Delta S°$ is positive – – –

63. a. 0 K   b. 1   c. greater

65. a. gases generally have high entropies compared to solids and

    b. $\Delta S°$ is the difference in entropy between products and reactants

    c. solid is more ordered

67. $K_a = \dfrac{(1.2 \times 10^{-5})^2}{0.200} = 7.2 \times 10^{-10}$; $1/K_a = 1.4 \times 10^{9}$

    $\Delta G° = -0.00831(298) \ln (1.4 \times 10^{9}) = -52.1$ kJ

69. a. $\Delta G° = -368.57$ kJ $- 711.6$ kJ $+ 16.5$ kJ $+ 101.4$ kJ

      $= -962.3$ kJ; yes

    b. yes

    c. no

71. $\Delta H° = (351.5)(0.2827 - 0.1607)$kJ $= 42.9$ kJ

    $\ln \dfrac{760}{357} = \dfrac{42,900(352 - T)}{8.31(352)T}$; $T = 335$ K $= 62°C$

174

72. $\Delta H° = 62.4$ kJ $- 53.0$ kJ $= +9.4$ kJ

$\Delta S° = 0.2607$ kJ/K $+ 0.1306$ kJ/K $- 0.4130$ kJ/K

$= -0.0217$ kJ/K

$\Delta G° = 9.4$ kJ $+ 773(0.0217)$kJ $= 26.2$ kJ

$\ln K = \dfrac{26.2}{(8.31 \times 10^{-3})(773)}$; $K = 0.017$

$\dfrac{(0.200 - x)^2}{(0.200 + 2x)^2} = 0.017$; $\quad \dfrac{0.200 - x}{0.200 + 2x} = 0.13$; $\quad x = 0.14$

$P_{H_2} = P_{I_2} = 0.06$ atm; $P_{HI} = 0.48$ atm

73. a. $(18.0)(0.333)$kJ $= +6.00$ kJ

    b. 0

    c. $\Delta S° = 6.00$ kJ/273 K $= 0.0220$ kJ/K

    d. $\Delta G° = +6.00$ kJ $- 5.56$ kJ $= +0.44$ kJ

    e. $\Delta G° = +6.00$ kJ $- 6.44$ kJ $= -0.44$ kJ

74. a. $\Delta S° = +140$ kJ/298 K $= +0.470$ kJ/K

    $\Delta G°$ at 37°C: -5650 kJ $- 310(0.470)$kJ $= -5800$ kJ

    $W = 1.00$ g $\times \dfrac{1\ \text{mol}}{342.3\ \text{g}} \times \dfrac{5800\ \text{kJ}}{1\ \text{mol}} \times 0.25 = 4.2$ kJ

    b. $W = (9.79 \times 10^{-3})(120/2.20)(4158) = 2.22 \times 10^3$ kJ

    mass $= \dfrac{2.22 \times 10^3\ \text{kJ}}{4.2\ \text{kJ/g}} = 5.3 \times 10^2$ g

75. $CaH_2(s) \longrightarrow Ca(s) + H_2(g)$

    $\Delta H° = +186.2$ kJ

    $\Delta S° = +0.0414$ kJ/K $+ 0.1306$ kJ/K $- 0.0420$ kJ/K $= +0.1300$ kJ/K

    $T = 186.2/0.1300 = 1430$ K $\approx 1160°C$

76. $2CuO(s) \longrightarrow Cu_2O(s) + \frac{1}{2} O_2(g)$

$\Delta H° = +146.0$ kJ; $\Delta S° = +0.1104$ kJ/K; $T = 1322$ K $\approx 1050°C$

$Cu_2O(s) \longrightarrow 2Cu(s) + \frac{1}{2} O_2(g)$

$\Delta H° = +168.6$ kJ; $\Delta S° = +0.0758$ kJ/K; $T = 2220$ K $\approx 1950°C$

CHAPTER 18
Electrochemistry

LECTURE NOTES

The most difficult portion of this chapter involves the Nernst equation. Students have a great deal of trouble using this equation to calculate the concentration of a species, knowing E and E°. The principal problem here is their unfamiliarity with logarithms. Other topics in this chapter, such as applications of E°, $\Delta G°$, and K, electrolysis calculations, etc., seem to go smoothly.

You should expect to spend at least 2½ lectures on this chapter, 3 if you want to discuss commercial cells in any detail.

<u>LECTURE 1</u>

I <u>Voltaic Cells</u>

Spontaneous reaction used to produce electrical energy.

A. <u>Salt bridge cells</u>

$$Zn(s) + Cu^{2+}(aq) \longrightarrow Zn^{2+}(aq) + Cu(s)$$

Must design cell to make electron transfer occur indirectly.

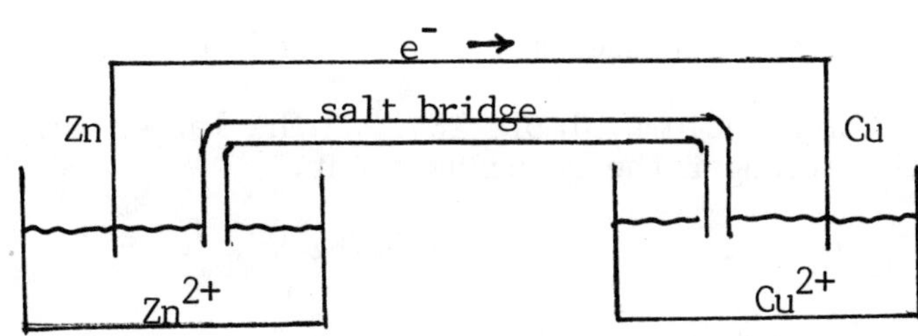

anode: $Zn(s) \longrightarrow Zn^{2+}(aq) + 2e^-$

cathode: $Cu^{2+}(aq) + 2e^- \longrightarrow Cu(s)$

Salt bridge allows current to flow but prevents contact between zinc and copper(II) ions, which would shortcircuit the cell.

For the cell:

$$Zn(s) + 2H^+(aq) \longrightarrow Zn^{2+}(aq) + H_2(g)$$

use an inert electrode, perhaps Pt, for the hydrogen half cell.

B. Cell Notation

$Zn - Cu^{2+}$ cell: $\quad Zn \mid Zn^{2+} \parallel Cu^{2+} \mid Cu$

$Zn - H^+$ cell $\quad : \quad Zn \mid Zn^{2+} \parallel H^+ \mid H_2 \mid Pt$

II Standard Voltages

A. E° = cell voltage when all species are at standard concentrations (1 atm for gases, 1 M for solutes in water)

$$Zn(s) + 2H^+(aq, 1\ M) \longrightarrow Zn^{2+}(aq, 1\ M) + H_2(g, 1\ atm)$$

$$E^\circ = +\ 0.762\ V = E^\circ_{ox}\ Zn + E^\circ_{red}\ H^+$$

arbitrarily set $E^\circ_{red}\ H^+ = 0.000\ V$; then $E^\circ_{ox} Zn = +0.762\ V$

Table 18.1 lists values of E°_{red}; can get E°_{ox} by changing sign

$$Cu(s) \longrightarrow Cu^{2+}(aq) + 2e^-; \quad E^\circ_{ox}\ Cu = -E^\circ_{red}\ Cu^{2+} = -0.339\ V$$

B. Relative strengths of oxidizing and reducing agents

1. Oxidizing agents (left column, Table 18.1). The larger (more positive) the value of E°_{red}, the stronger the oxidizing agent.

$$Zn^{2+}(aq) + 2e^- \longrightarrow Zn(s) \qquad E^\circ_{red} = -0.762\ V$$

$$2H^+(aq) + 2e^- \longrightarrow H_2(g) \qquad E^\circ_{red} = 0.000\ V$$

$$Cl_2(g) + 2e^- \longrightarrow 2Cl^-(aq) \qquad E^\circ_{red} = -1.360\ V$$

oxidizing agents become stronger moving down left column

2. Reducing agents (right column, Table 18.1). The larger the value of E°_{ox}, the stronger the reducing agent.

$$Zn(s) \longrightarrow Zn^{2+}(aq) + 2e^- \qquad E^\circ_{ox} = +0.762\ V$$

$$H_2(g) \longrightarrow 2H^+(aq) + 2e^- \qquad E^\circ_{ox} = 0.000\ V$$

$$2Cl^-(aq) \longrightarrow Cl_2(g) + 2e^- \qquad E^\circ_{ox} = -1.360\ V$$

reducing agents become weaker moving down right column

C. <u>Calculation of $E°$</u> $E° = E°_{ox} + E°_{red}$

$$Cl_2(g) + 2Br^-(aq) \longrightarrow 2Cl^-(aq) + Br_2(l)$$

$$E° = E°_{red} \; Cl_2 + E°_{ox} \; Br^- = 1.360 \; V - 1.077 \; V = +0.283 \; V$$

Since calculated voltage is positive, this reaction can occur in a voltaic cell.

D. <u>Determination of whether redox reaction will occur</u>

What, if anything, happens when bromine is added to a solution of tin(II) chloride?

Possible oxidations:

$$Sn^{2+}(aq) \longrightarrow Sn^{4+}(aq) + 2e^- \qquad E°_{ox} = -0.154 \; V$$

$$2Cl^-(aq) \longrightarrow Cl_2(g) + 2e^- \qquad E°_{ox} = -1.360 \; V$$

Possible reductions:

$$Sn^{2+}(aq) + 2e^- \longrightarrow Sn(s) \qquad E°_{red} = -0.141 \; V$$

$$Br_2(l) + 2e^- \longrightarrow 2Br^-(aq) \qquad E°_{red} = +1.077 \; V$$

Reaction:

$$Sn^{2+}(aq) + Br_2(l) \longrightarrow Sn^{4+}(aq) + 2Br^-(aq); \; E° = +0.923 \; V$$

LECTURE 2

I <u>Relation Between $E°$, $\Delta G°$, and K</u>

$$\Delta G° = -nFE° \qquad\qquad \ln K = nE°/0.0257$$

Note that if $E°$ is +, $\Delta G°$ is −, $\ln K$ is +, $K > 1$

$$Cl_2(g) + 2Br^-(aq) \longrightarrow 2Cl^-(aq) + Br_2(l)$$

$$E° = 1.360 \; V - 1.077 \; V = +0.283 \; V$$

$$\Delta G° = -2(96.5)(+0.283)kJ = -54.6 \; kJ$$

$$\ln K = \frac{2(0.283)}{0.0257} = 22.0; \; K = 4 \times 10^9$$

II <u>Effect of Concentration upon Voltage</u>

A. <u>Nernst Equation</u>

$$E = E^{\circ} + \frac{RT}{nF} \ln Q = E^{\circ} - \frac{0.0257}{n} \ln Q$$

In expression for Q, gases enter as partial pressures in atmospheres, solutes as molarities.

$$Cl_2(g) + 2Br^-(aq) \longrightarrow 2Cl^-(aq) + Br_2(l)$$

$$E = +0.283 \text{ V} - \frac{0.0257}{2} \ln \frac{[Cl^-]^2}{(P\,Cl_2) \times [Br^-]^2}$$

Calculate voltage when $[Br^-] = 1$ M, $P\,Cl_2 = 1$ atm, $[Cl^-] = 0.01$ M

$$E = +0.283 \text{ V} - \frac{0.0257}{2} \ln(0.01)^2 = +0.401 \text{ V}$$

B. Use of Nernst equation to determine concentrations of ions in solution

$$Zn(s) + 2H^+(aq) \longrightarrow Zn^{2+}(aq) + H_2(g)$$

$$E = +0.762 \text{ V} - \frac{0.0257}{2} \ln \frac{(P\,H_2) \times [Zn^{2+}]}{[H^+]^2}$$

Suppose $[Zn^{2+}] = 1$ M, $P\,H_2 = 1$ atm

$$E = +0.762 \text{ V} + 0.0257 \ln [H^+]$$

Measure voltage, calculate $[H^+]$. Suppose $E = 0.200$ V

$$\ln [H^+] = \frac{-0.562}{0.0257} = -21.9; \quad [H^+] = 3 \times 10^{-10} \text{ M}$$

LECTURE 2½

I <u>Electrolytic Cells</u>

Electrical energy supplied to bring about a nonspontaneous redox reaction. Cell diagram.

A. Amount of products formed

    1. $Ag^+(aq) + e^- \longrightarrow Ag(s)$ 1 mol $e^- = 96480$ C $\longrightarrow$ 1 mol Ag

        no. of coulombs = no. of amperes x no. of seconds

        no. of joules = no. of coulombs x no. of volts

180

2. How much silver is plated from $AgNO_3$ solution by a current of 2.60 A in one hour?

$$n \ e^- = (2.60)(3600)C \ \times \ \frac{1 \ mol \ e^-}{96480 \ C} = 0.0970 \ mol \ e^-$$

$$mass \ Ag = 0.0970 \ mol \ e^- \ \times \ \frac{1 \ mol \ Ag}{1 \ mol \ e^-} \ \times \ \frac{107.9 \ g \ Ag}{1 \ mol \ Ag} = 10.5 \ g \ Ag$$

II <u>Commercial cells</u>

A. Electrolysis of aqueous NaCl

$$2H_2O + 2Cl^-(aq) \longrightarrow H_2(g) + Cl_2(g) + 2 \ OH^-(aq)$$

Voltage required = 1.360 V + 0.828 V = 2.188 V

Energy required to form one mole of Cl_2?

$$no. \ coulombs = 1 \ mol \ Cl_2 \ \times \ \frac{2 \ mol \ e^-}{1 \ mol \ Cl_2} \ \times \ \frac{96480 \ C}{1 \ mol \ e^-} = 1.930 \times 10^5 \ C$$

$$no. \ joules = (1.930 \times 10^5)(2.188) = 4.223 \times 10^5 \ J = 422.3 \ kJ$$

B. Lead storage battery

anode: $\quad Pb(s) + SO_4^{2-}(aq) \longrightarrow PbSO_4(s) + 2e^-$

cathode: $PbO_2(s) + 4H^+(aq) + SO_4^{2-}(aq) + 2e^- \longrightarrow PbSO_4(s) + 2H_2O$

overall: $Pb(s) + PbO_2(s) + 4H^+(aq) + 2SO_4^{2-}(aq) \longrightarrow 2PbSO_4(s) + 2H_2O$

As cell discharges, conc. H_2SO_4 and density decrease

DEMONSTRATIONS

1. Electricity by chemical reactions: GILB G 30

2. Electron flow in redox reactions: GILB G 16

3. Human salt bridge: J. Chem. Educ. <u>67</u> 156 (1990)

4. The voltaic pile: J. Chem. Educ. <u>68</u> 665 (1991)

5. Half-cell reactions: J. Chem. Educ. <u>68</u> 247 (1991)

6. Vegetable voltage and fruit juice: GILB G 37

Saunders College Publishing

7. Electrode potentials: GILB G 43

8. Hydrogen electrode: GILB G 45

9. Zn - H^+ cell: SHAK $\underline{4}$ 130

10. Electrodeposition of nickel on copper: GILB G 56

11. Reactions of metals with HCl: SHAK $\underline{1}$ 25; J. Chem. Educ. $\underline{72}$ 55 (1995)

12. Concentration cells: GILB G 32; SHAK $\underline{4}$ 140

13. Electroplating: GILB E 4; SHAK $\underline{4}$ 212-244

14. Electrolysis of water solutions: SHAK $\underline{4}$ 170-181, 205-209

15. Commercial voltaic cells: SHAK $\underline{4}$ 115; J. Chem. Educ. $\underline{67}$ 158 (1990)

16. Corrosion: GILB G 63, M 261

PROBLEMS

1. a. $Cu(s) + I_2(s) \longrightarrow Cu^{2+}(aq) + 2I^-(aq)$

   b. $Mn^{2+}(aq) + Cl_2(g) + 2H_2O \longrightarrow MnO_2(s) + 2Cl^-(aq) + 4H^+(aq)$

   c. $2Al(s) + 3Fe^{2+}(aq) \longrightarrow 2Al^{3+}(aq) + 3Fe(s)$

3. a. Pb anode, Pt cathode; e^- flow from Pb to Pt; anions move to Pb, cations to Pt.

   b. Au anode, Pt cathode; e^- flow from Au to Pt; anions move to Au, cations to Pt.

   c. Fe anode, Pt cathode; e^- flow from Fe to Pt; anions move to Fe, cations to Pt.

5. $Fe(s) + Cu(OH)_2(s) \longrightarrow Fe(OH)_2(s) + Cu(s)$

   $Fe \mid Fe(OH)_2 \parallel Cu(OH)_2 \mid Cu$

   e^- flow from Fe to Cu; anions move to Fe, cations to Cu

7. a. NO_3^- b. MnO_2 c. S

9. Cr^{3+} (E°_{ox} = -1.33 V) < NO acidic (E°_{ox} = -0.964 V) < H_2 acidic (E°_{ox}

   = 0.000 V) < NO basic (E°_{ox} = +0.140 V) < Cd (E°_{ox} = +0.402 V)

Saunders College Publishing

11. oxid. agents: Cr^{3+}, $Cr_2O_7^{2-}$, H_2O (basic)

$$H_2O(\text{basic}) < Cr^{3+} < Cr_2O_7^{2-}$$

red. agents: Cr^{2+}, Cr^{3+}, Cr

$$Cr^{3+} < Cr^{2+} < Cr$$

13. a. Ni or Ag + I^-

b. Hg, Fe^{2+}, I^-

c. $Cr_2O_7^{2-}$, Cl_2, ClO_3^-, Au^{3+}

15. a. $E° = +0.141$ V $+ 0.799$ V $= +0.940$ V

b. $E° = +0.000$ V $+ 0.796$ V $= +0.796$ V

c. $E° = +0.356$ V $+ 1.687$ V $= +2.043$ V

17. a. $E° = 0.000$ V $+ 0.769$ V $= +0.769$ V

b. $E° = -1.360$ V $+ 1.512$ V $= +0.152$ V

c. $E° = +0.828$ V $- 0.140$ V $= +0.688$ V

19. a. $E° = +0.282$ V $+ 0.799$ V $= +1.081$ V

b. $E° = +0.762$ V $+ 1.077$ V $= +1.839$ V

c. $E° = -0.534$ V $+ 1.001$ V $= +0.467$ V

21. a. -0.236 V   b. $+0.173$ V   c. $+1.081$ V; same

23. a. $E° = -1.229$ V $+ 1.360$ V $= +0.131$ V; spontaneous

b. $E° = -0.762$ V $- 0.769$ V $= -1.531$ V; nonspontaneous

c. $E° = +1.077$ V $- 0.534$ V $= +0.543$ V; spontaneous

25. $E°_{ox} = -1.458$ V;   a. yes   b. no   c. yes

27. a. no reaction

b. $6Hg(1) + 2NO_3^-(aq) + 8H^+(aq) \longrightarrow 3Hg_2^{2+}(aq) + 2NO(g) + 4H_2O$

$E° = -0.796$ V $+ 0.964$ V $= +0.168$ V

c. no reaction

29. a, d

31. a. E°_{ox} Cr = +0.744 V; E°_{red} Ni^{2+} = -0.236 V

$$2Cr(s) + 3Ni^{2+}(aq) \longrightarrow 2Cr^{3+}(aq) + 3Ni(s)$$

b. $F_2(g) + 2Cl^-(aq) \longrightarrow 2F^-(aq) + Cl_2(g)$; E° = +1.529 V

c. E°_{red} S = +0.144 V; E°_{red} Fe^{3+} = +0.769 V; NO_3^- E°_{red} = +0.964 V

no reaction (no oxidation)

33. a. E° = -0.025 V; K = 7.9 x 10^{-3}

b. ΔG° = -108 kJ; K = 7.3 x 10^{18}

c. ΔG° = -27 kJ; E° = 0.055 V

35. ΔG° = -2326.0 kJ + 684.3 kJ + 474.4 kJ + 894.4 kJ = -272.9 kJ

ΔE° = +272,900/6(96480) = +0.4714 V

K = 7.3 x 10^{47}

37. a. ΔG° = -2(96,480)(0.940)J = -181 kJ

b. ΔG° = -2(96480)(0.796)J = -154 kJ

c. ΔG° = -2(96480)(2.043)J = -394.2 kJ

39. a. $H_2(g) + 2Fe^{3+}(aq) \longrightarrow 2H^+(aq) + 2Fe2+(aq)$

ln K = $\dfrac{2(0.769)}{0.0257}$; K = 1 x 10^{26}

b. $2MnO_4^-(aq) + 10Cl^-(aq) + 16H^+(aq) \longrightarrow 2Mn^{2+}(aq) + 5Cl_2(g) + 8H_2O$

ln K = $\dfrac{10(0.152)}{0.0257}$; K = 5 x 10^{25}

c. $3H_2(g) + 2NO_3^-(aq) \longrightarrow 2NO(g) + 2 OH^-(aq) + 2H_2O$

ln K = $\dfrac{6(0.688)}{0.0257}$; K = 6 x 10^{69}

41. a. E° = +1.68 V

b. E = 1.68 - $\dfrac{0.0591}{6}$ $\log_{10} \dfrac{(P\ H_2)^3 \times [Al^{3+}]^2}{[H^+]^6}$

184

Saunders College Publishing

c. $E = 1.68$ V $- \dfrac{0.0591}{6} \log_{10} \dfrac{(1.00)(1.00 \times 10^{-2})}{1 \times 10^{-18}} = 1.52$ V

43. a. $E° = -0.769$ V $+ 1.763$ V $= 0.994$ V

b. $E = +0.994$V$- \dfrac{0.0591}{2} \log_{10} \dfrac{[Fe^{3+}]^2}{[H_2O_2] \times [Fe^{2+}]^2 \times [H^+]^2}$

c. $E = +0.994$ V $- \dfrac{0.0591}{2} \log_{10} \dfrac{(0.214)^2}{(1.50)(0.0037)^2(6.9 \times 10^{-4})^2}$

$= +0.708$ V

45. $2Cr(s) + 3Co^{2+}(aq) \longrightarrow 2Cr^{3+}(aq) + 3Co(s)$

$E° = +0.744$ V $- 0.282$ V $= +0.462$ V

$E = +0.462$ V $- \dfrac{0.0591}{6} \log_{10} \dfrac{(2.0 \times 10^{-3})^2}{(1.5)^3} = 0.520$ V

47. $E° = -1.001$ V $+ 1.077$ V $= +0.076$ V

$0.000 = 0.076 - \dfrac{0.0591}{6} \log_{10} \dfrac{(0.100)^8}{[Cl^-]^8}$

Solving, $[Cl^-] = 0.011$ M

49. $Cu(s) + 2Ag^+(aq) \longrightarrow Cu^{2+}(aq) + 2Ag(s)$

$+0.15$ V $= 0.460$ V $- \dfrac{0.0591}{2} \log_{10} \dfrac{0.020}{[Ag^+]^2}$

$[Ag^+] = 8 \times 10^{-7}$ M

51. $E° = -1.229$ V $+ 1.077$ V $= -0.152$ V

$E = -0.152$ V $- \dfrac{0.0591}{4} \log_{10} (1.0 \times 10^{-16})$

$= -0.152$ V $+ 0.236$ V $= +0.084$ V (spontaneous)

53. a. $E° = -0.799$ V $+ 0.339$ V $= -0.460$ V

b. $0.060 = -0.460 - \dfrac{0.0591}{2} \log_{10}[Ag^+]^2$

$0.520 = -0.0591 \log_{10} [Ag^+]; \quad [Ag^+] = 1.6 \times 10^{-9}$ M

185

Saunders College Publishing

c. $K_{sp} = 1.6 \times 10^{-10}$

55. a. $n\ Al = \dfrac{5000}{26.98}$; $n\ e^- = \dfrac{3(5000)}{26.98} = 556.0$ mol e^-

b. no. of amperes $= \dfrac{556.0 \times 96480}{24 \times 3600} = 620.9$ A

c. $Al_2O_3(s) \longrightarrow 2Al(s) + 3/2\ O_2(g)$

$n\ O_2 = \dfrac{5000}{26.98} \times \dfrac{3/2}{2} = 139.0$ mol

57. a. area $= (2.5 \times 10.0)2 + (2.5 \times 0.0500)2 + (10.0 \times 0.0500)2$

$= 51.25\ in^2$

mass Ag $= 0.05125\ in^3 \times (2.54\ cm/1\ in)^3 \times 10.5\ g/cm^3 = 8.82$ g

b. no. of coulombs $= \dfrac{8.82 \times 96480}{107.9} = (4.00)t$

$t = 1.97 \times 10^3$ s $= 0.547$ hr

59. a. $\dfrac{(2.00)(3600)}{96480}$ mol $e^- \times \dfrac{1\ mol\ Pb}{2\ mol\ e^-} \times \dfrac{207.2\ g\ Pb}{1\ mol\ Pb} = 7.73$ g Pb

b. $\dfrac{(2.00)(3600)(12.0)}{3.600 \times 10^6} = 0.0240$ kWh

61. a. allows current to flow without bringing reactants into direct contact

b. cations always move to cathode

c. H_2SO_4 participates in cell reaction

63. $Co(s) + 2H^+(aq) \longrightarrow Co^{2+}(aq) + H_2(g)$

only (b)

65. $n\ H_2 = \dfrac{(5.00)(1.20)}{(0.0821)(298)}$; $n\ e^- = \dfrac{(5.00)(1.20)2}{(0.0821)(298)}$

no. coulombs $= \dfrac{(5.00)(1.20)(2)(96480)}{(0.0821)(298)} = 10.0\ t$

$t = 4.73 \times 10^3$ s $= 1.31$ hr

67. $E = 0.152 \text{ V} - \dfrac{0.0591}{4} \log_{10} \dfrac{1}{(0.100)^4 (1.8 \times 10^{-5})^4} = -0.188 \text{ V}$

69. $\Delta H° = -393.5 \text{ kJ} + 110.5 \text{ kJ} = -283.0 \text{ kJ}$

    $\Delta S° = 0.2136 \text{ kJ/K} - 0.1976 \text{ kJ/K} - 0.1025 \text{ kJ/K} = -0.0865 \text{ kJ/K}$

    $\Delta G° = -283.0 \text{ kJ} + 1273(0.0865)\text{kJ} = -173 \text{ kJ}$

    $n = 2$

    $E° = \dfrac{173,000}{2(96480)} = 0.897 \text{ V}$

71. $Pb(s) + PbO_2(s) + 4H^+(aq) + 2SO_4{}^{2-}(aq) \longrightarrow 2PbSO_4(s) + 2H_2O$

Consider 1000 g of solution:

$n \; H_2SO_4 = 380 \text{ g} \times \dfrac{1 \text{ mol}}{98.08 \text{ g}} \; ; \quad V = 1000 \text{ g} \times \dfrac{1 \text{ cm}^3}{1.286 \text{ g}} \times \dfrac{1 \text{ L}}{1000 \text{ cm}^3}$

$[H_2SO_4] = 4.98 \text{ mol/L}$

$[H^+] = 4.98 \text{ M}; \quad [SO_4{}^{2-}] = 1.0 \times 10^{-2} \times \dfrac{[HSO_4{}^-]}{[H^+]} = 1.0 \times 10^{-2} \text{ M}$

$E° = 1.687 \text{ V} + 0.356 \text{ V} = 2.043 \text{ V}$

$E = 2.043 \text{ V} - \dfrac{0.0591}{2} \log_{10} \dfrac{1}{(4.98)^4 \times (1.0 \times 10^{-2})^2}$

   $= 2.043 \text{ V} - 0.036 \text{ V} = 2.007 \text{ V}$

72. a. $E° = 0.621 \text{ V}$

  b. $[Zn^{2+}]$ increases, $[Sn^{2+}]$ decreases

  c. $0 = 0.621 - \dfrac{0.0591}{2} \log_{10} \text{ratio}; \quad \text{ratio} = 1 \times 10^{21}$

  d. $[Zn^{2+}] = 2.0 \text{ M}; \quad [Sn^{2+}] = 2 \times 10^{-21} \text{ M}$

73.   a. $\Delta G^\circ = 2(9.648 \times 10^4)(0.581)\text{J} = 1.12 \times 10^5$ J

        $\Delta G^\circ = 2(9.648 \times 10^4)(0.197)\text{J} = 0.38 \times 10^5$ J

        $\Delta G^\circ_{tot} = 1.50 \times 10^5$ J

  b. $E^{\circ\prime} = -1.50 \times 10^5 / (4 \times 9.648 \times 10^4) = -0.389$ V

74.   anode: $H_2(g) \longrightarrow 2H^+(aq) + 2e^-$

     $E_{ox} = 0.00 \text{ V} - \dfrac{0.0591}{2} \log_{10}(1 \times 10^{-7})^2 = +0.414$ V

     cathode: $2H^+(aq) + 2e^- \longrightarrow H_2(g)$

     $E_{red} = 0; \; E = +0.414$ V

Saunders College Publishing

CHAPTER 19
Nuclear Reactions

LECTURE NOTES

This material typically is covered near the end of the school year. It serves to hold the interest of restless students. Nore than any other topic covered in general chemistry, nuclear reactions relate directly to issues that are meaningful to students.

The principles presented here are readily covered in $1\frac{1}{2}$ lectures; recall that radioactivity was discussed in Chapters 2 and 11. If more time is available, it can profitably be spent in a more detailed discussion of such topics as the effect of radiation on human beings, the pros and cons of nuclear reactors, and the prospects for fusion reactors.

<u>LECTURE 1</u>

I <u>Radioactivity</u>

  A. <u>Natural</u> Recall from Chapter 2 that alpha, beta, or gamma radiation can be given off.

  B. <u>Induced</u> Stable nucleus is bombarded by high energy particle (neutron, alpha, etc.). Unstable, radioactive nucleus is formed.

$$^{27}_{13}\text{Al} + \,^{1}_{0}\text{n} \longrightarrow \,^{28}_{13}\text{Al} \longrightarrow \,^{28}_{14}\text{Si} + \,^{0}_{-1}\text{e}$$

If isotope formed has too few neutrons, can get positron decay or K-electron capture.

$$^{11}_{6}\text{C} \longrightarrow \,^{11}_{5}\text{B} + \,^{0}_{1}\text{e} \qquad \text{or} \qquad ^{11}_{6}\text{C} + \,^{0}_{-1}\text{e} \longrightarrow \,^{11}_{5}\text{B}$$

II <u>Rate of Decay</u>

Recall (Chapter 11) that rate is first order

$$\ln X_o/X = kt; \quad k = 0.693/t_{\frac{1}{2}}$$

A. <u>Activity</u>

$$\text{activity} = kN$$

where N = number of atoms in sample, k = rate constant. Activity can be expressed in atoms decaying per unit time or in curies:

$$1 \text{ Ci} = 3.700 \times 10^{10} \text{ atoms decaying per second}$$

For radium, $k = 1.37 \times 10^{-11}/s$. Calculate the activity of a one milligram sample of radium.

$$N = 1.00 \times 10^{-3} \text{ g} \times \frac{1 \text{ mol}}{226 \text{ g}} \times 6.022 \times 10^{23} \text{ atoms/mol}$$

$$= 2.66 \times 10^{18} \text{ atoms}$$

$$\text{rate} = 1.37 \times 10^{-11}/s \times 2.66 \times 10^{18} \text{ atoms} = 3.64 \times 10^{7} \text{ atoms/s}$$

$$\approx 1 \text{ millicurie}$$

B. <u>Age of organic material</u>; measure C-14 content

$$^{14}_{7}\text{N} + ^{1}_{0}\text{n} \longrightarrow ^{14}_{6}\text{C} + ^{1}_{1}\text{H}$$

$$^{14}_{6}\text{C} \longrightarrow ^{14}_{7}\text{N} + ^{0}_{-1}\text{e} \; ; \; t_{\frac{1}{2}} = 5720 \text{ yr}$$

In a living plant or animal, these two processes are in equilibrium, C-14 content is constant. When plant or animal dies, first process stops, C-14 content declines.

Suppose fragment of Shroud of Turin showed C-14 content 0.930 times that of living plant. Age of shroud?

$$k = 0.693/5720 \text{ yr} = 1.21 \times 10^{-4} \text{ yr}$$

$$\ln 1/0.930 = (1.21 \times 10^{-4})t; \quad t = 600 \text{ yr}$$

III <u>Mass-Energy Relations</u>

A. Relation between ΔE and m; $\Delta E = c^{2}\Delta m$

$$\Delta E \text{ (in joules)} = (3.0 \times 10^{8})^{2} \Delta m(\text{in kilograms})$$

$$\Delta E \text{ (in kilojoules)} = 9.0 \times 10^{10} \Delta m \text{ (in grams)}$$

B. Calculations

$$^{239}_{94}\text{Pu} \longrightarrow ^{4}_{2}\text{He} + ^{235}_{92}\text{U}$$

190

$$\Delta m \text{ per mole} = 234.9934 \text{ g} + 4.0015 \text{ g} - 239.0006 \text{ g} = -0.0057 \text{ g}$$

$$\Delta E \text{ per mole} = -0.0057 \text{ g} \times 9.0 \times 10^{10} \text{ kJ/g} = -5.1 \times 10^8 \text{ kJ}$$

$$\Delta E \text{ per gram} = \frac{-5.1 \times 10^8 \text{ kJ}}{239 \text{ g}} = -2.1 \times 10^6 \text{ kJ/g}$$

This compares to a maximum of about 50 kJ/g for ordinary chemical reactions.

LECTURE 1½

I **Mass Defect** = mass (n + p) − mass nucleus

$^{12}_{6}$C: $6(1.00867 \text{ g/mol} + 1.00728 \text{ g/mol}) - 11.99671 \text{ g/mol} = 0.09899 \text{ g/mol}$

Binding energy = 9.00×10^{10} kJ/g $\times$ 0.09899 g/mol = 8.91×10^9 kJ/mol

Plot of binding energy/(no. of nucleons) vs mass number gives maximum at 50-90. Fission of very heavy nucleus or fusion of very light nuclei evolves energy.

II **Fission**

$$^{235}_{92}\text{U} + ^{1}_{0}\text{n} \longrightarrow ^{90}_{37}\text{Rb} + ^{144}_{55}\text{Cs} + 2\,^{1}_{0}\text{n}$$

Note that:

1. Many different isotopes are formed.
2. More neutrons are produced than consumed, leading to a chain reaction. In nuclear reactor, excess neutrons are absorbed by cadmium rods.
3. Nuclei produced have too many neutrons and hence are intensely radioactive:

$$^{90}_{37}\text{Rb} \longrightarrow ^{0}_{-1}\text{e} + ^{90}_{38}\text{Sr}$$

This is the principal danger associated with nuclear reactors. Three Mile Island and Chernobyl.

III **Fusion**

$$2\,^{2}_{1}\text{H} \longrightarrow ^{4}_{2}\text{He}; \quad \Delta m = -0.02560 \text{ g}; \quad \Delta E = -2.3 \times 10^9 \text{ kJ}$$

$$\text{per gram:} \quad \Delta E = \frac{-2.3 \times 10^9 \text{ kJ}}{4.0 \text{ g}} = -5.7 \times 10^8 \text{ kJ/g}$$

ΔE per gram is about 10 times that for fission, 200 times that for radioactivity. Unfortunately, activation energy is very high, since

191

two deuterons repel each other. Temperature required is of the order
a billion °C.

DEMONSTRATIONS

1. Absorption of beta rays: GILB L 46

2. Cloud chamber: GILB L 45

<u>VIDEO</u>

1. Shielding radiation: Falcon 10

2. Radiation from common sources: Falcon 11

3. Simulation of nuclear chain reactions: SHAK 49

PROBLEMS

1. $${}^{282}_{115}X + {}^{0}_{-1}e \longrightarrow {}^{282}_{114}Y$$

$${}^{282}_{115}X \longrightarrow {}^{0}_{1}e + {}^{282}_{114}Y$$

products are the same

3. $${}^{51}_{24}Cr \longrightarrow {}^{0}_{1}e + {}^{51}_{23}V$$

5. a. $$2\,{}^{12}_{6}C \longrightarrow {}^{1}_{0}n + {}^{23}_{12}Mg$$

b. $${}^{235}_{92}U + {}^{1}_{0}n \longrightarrow {}^{140}_{56}Ba + {}^{93}_{36}Kr + 3\,{}^{1}_{0}n$$

c. $${}^{210}_{88}Ra \longrightarrow {}^{4}_{2}He + {}^{206}_{86}Rn$$

7. a. $${}^{109}_{47}Ag + {}^{4}_{2}He \longrightarrow {}^{113}_{49}In$$

b. $${}^{246}_{96}Cm + {}^{12}_{6}C \longrightarrow 4\,{}^{1}_{0}n + {}^{254}_{102}No$$

c. $${}^{96}_{42}Mo + {}^{2}_{1}H \longrightarrow {}^{1}_{0}n + {}^{97}_{43}Tc$$

d. $^{31}_{16}\text{S} + ^{1}_{0}\text{n} \longrightarrow ^{1}_{1}\text{H} + ^{31}_{15}\text{P}$

9. a. $^{126}_{56}\text{Ba} + ^{0}_{1}\text{e} \longrightarrow ^{126}_{57}\text{La}$

b. $^{125}_{56}\text{Ba} \longrightarrow ^{0}_{1}\text{e} + ^{125}_{55}\text{Cs}$

c. $^{14}_{7}\text{N} + ^{4}_{2}\text{He} \longrightarrow ^{17}_{8}\text{O} + ^{1}_{1}\text{H}$

d. $^{24}_{12}\text{Mg} + ^{4}_{2}\text{He} \longrightarrow ^{27}_{14}\text{Si} + ^{1}_{0}\text{n}$

11. $2156 \text{ Ci} \times \dfrac{3.700 \times 10^{10}}{\text{s}} \times \dfrac{60 \text{ s}}{1 \text{ min}} = 4.786 \times 10^{15}/\text{min}$

13. $\dfrac{2.565 \times 10^{4}}{10 \text{ min}} \times \dfrac{1 \text{ min}}{60 \text{ s}} = 42.75/\text{s} = 1.155 \times 10^{-9} \text{ Ci}$

15. $\text{activity} = \dfrac{1.5 \times 10^{-4}}{\text{s}} \times \dfrac{10.0 \times 10^{-3}}{87.0} \times 6.022 \times 10^{23} = 1.0 \times 10^{16} \text{ atom/s}$

17. $k = \dfrac{0.693}{36 \text{ hr}} \times \dfrac{1 \text{ hr}}{60 \text{ min}} = 3.2 \times 10^{-4}/\text{min}$

$\dfrac{8.7 \times 10^{4}}{\text{min}} = \dfrac{3.2 \times 10^{-4}}{\text{min}} \times (\text{no. of atoms}); \quad \text{no. of atoms} = 2.7 \times 10^{8}$

$\text{mass} = 2.7 \times 10^{8} \text{ atoms} \times \dfrac{82.0 \text{ g}}{6.022 \times 10^{23} \text{ atoms}} = 3.7 \times 10^{-14} \text{ g}$

19. $k = \dfrac{0.693}{6.0 \text{ hr}} \times \dfrac{1 \text{ hr}}{3600 \text{ s}} = 3.2 \times 10^{-5}/\text{s}$

$\text{activity} = \dfrac{3.2 \times 10^{-5}}{\text{s}} \times \dfrac{1.00 \times 10^{-3}}{98.9} \times 6.022 \times 10^{23} = 2.0 \times 10^{14}/\text{s}$

21. $k = \dfrac{2.3 \times 10^{-6}}{\text{yr}} \times \dfrac{1 \text{ yr}}{365 \text{ d}} \times \dfrac{1 \text{ d}}{24 \text{ hr}} \times \dfrac{1 \text{ hr}}{3600 \text{ s}} = 7.3 \times 10^{-14}/\text{s}$

$\text{activity} = \dfrac{7.3 \times 10^{-14}}{\text{s}} \times \dfrac{10.0 \times 10^{-3}}{36.0} \times 6.022 \times 10^{23} = 1.2 \times 10^{7}/\text{s}$

$\text{counts per } \tfrac{1}{2} \text{ min} = 30(1.2 \times 10^{7}) = 3.6 \times 10^{8}$

$\text{no. of curies} = \dfrac{1.2 \times 10^{7}}{\text{s}} \times \dfrac{1 \text{ Ci}}{3.700 \times 10^{10}/\text{s}} = 3.2 \times 10^{-4} \text{ Ci}$

23. $\ln \dfrac{22.1}{12.0} = \dfrac{0.693}{5720 \text{ yr}} \times t$; $t = 5.04 \times 10^3$ yr

25. $\ln \dfrac{1.00}{0.62} = \dfrac{0.693}{12.3 \text{ yr}} \times t$; $t = 8.5$ yr

27. mole ratio $= \dfrac{1.30/238}{1.00/206} = 1.13$

$\ln \dfrac{2.13}{1.13} = \dfrac{0.\,693\,t}{(4.5 \times 10^9 \text{ yr})}$; $t = 4.1 \times 10^9$ yr

29. Ar $-$ K: $\ln 5.13 = 0.693t/(1.26 \times 10^9 \text{ yr})$ $t = 2.97 \times 10^9$ yr

Pb $-$ U: $\ln 1.66 = 0.693t/(4.5 \times 10^9 \text{ yr})$ $t = 3.3 \times 10^9$ yr

Sr $-$ Rb: $\ln 1.049 = 0.693t/(4.8 \times 10^{10} \text{ yr})$ $t = 3.3 \times 10^9$ yr

loss of Ar(g)

31. $^{222}_{86}\text{Rn} \longrightarrow \, ^{4}_{2}\text{He} + \, ^{218}_{84}\text{Po}$

a. $\Delta m = 217.9628$ g $+ 4.0015$ g $- 221.9703$ g $= -0.0060$ g

b. per mole, $\Delta E = -9.00 \times 10^{10}$ kJ/g $\times 0.0060$ g $= -5.4 \times 10^8$ kJ

per gram: $\Delta E = \dfrac{-5.4 \times 10^8 \text{ kJ}}{222 \text{ g}} = -2.4 \times 10^6$ kJ/g

33. a. $97(1.00728$ g$) + 148(1.00867$ g$) - 245.1029$ g $= 1.8864$ g

b. 9.00×10^{10} kJ/g $\times 1.8864$ g $= 1.70 \times 10^{11}$ kJ

35. $^{19}_{9}\text{F}$: $9(1.00728\text{g}) + 10(1.00867$ g$) - 18.99346$ g $= 0.15876$ g

$^{17}_{8}\text{O}$: $8(1.00728$ g$) + 9(1.00867$ g$) - 16.99474$ g $= 0.14153$ g

F-19 is larger

37. $\text{N}_2\text{H}_4(1) + \text{O}_2(g) \longrightarrow \text{N}_2(g) + 2\text{H}_2\text{O}(g)$

$\Delta H = -483.6$ kJ $- 50.6$ kJ $= -534.2$ kJ

$\Delta m = -534.2$ kJ/9.00×10^{10} kJ/g $= -5.94 \times 10^{-9}$ g

194

Saunders College Publishing

39. Δm/mol = 88.8913 g + 143.8817 g + 2(1.00867 g) + 3(0.00055 g)

$$- 234.9934 \text{ g} = -0.2014 \text{ g}$$

a. ΔE = -9.00 x 10^{10} kJ x $\dfrac{0.2014}{235.0}$ = -7.71 x 10^7 kJ

b. $\dfrac{7.71 \times 10^7 \text{ kJ}}{2.76 \text{ kJ/g}}$ = 2.79 x 10^7 g = 2.79 x 10^4 kg

c. $\dfrac{2.76 \times 10^6 \text{ kJ}}{7.71 \times 10^7 \text{ kJ/g}}$ = 3.58 x 10^{-2} g

41. $^{26}_{13}\text{Al} + ^{0}_{-1}\text{e} \longrightarrow ^{26}_{12}\text{Mg}$

Δm = 25.97600 g - 0.00055 g - 25.97977 g = 0.00432 g; spontaneous

43. a. $^{235}_{92}\text{U} + ^{1}_{0}\text{n} \longrightarrow ^{144}_{58}\text{Ce} + ^{87}_{35}\text{Br} + ^{0}_{-1}\text{e} + 5\,^{1}_{0}\text{n}$

b. Δm per mole = 143.8817 g + 86.9028 g + 0.00055 g + 4(1.00867 g)

$$- 234.9934 \text{ g} = -0.1737 \text{ g}$$

ΔE per gram = -9.00 x 10^{10} x $\dfrac{0.1737}{235.0}$ = -6.65 x 10^7 kJ

c. 37.0 $\dfrac{\text{kJ}}{\text{mol}}$ x $\dfrac{1 \text{ mol}}{80.0 \text{ g}}$ x $\dfrac{10^3 \text{ g}}{1 \text{ kg}}$ = 4.62 x 10^2 kJ/kg

$\dfrac{6.65 \times 10^4 \text{ kg}}{4.62 \times 10^2}$ = 1.44 x 10^2 kg

45. a. mass number doesn't change; atomic number increases

b. decays very rapidly

c. more energy

47. ln $\dfrac{15.0}{x}$ = $\dfrac{0.693}{8.1}$ x 30 = 2.6

15.0/x = 13.0; 1.2 mg

49. a. ΔE = -9.00 x 10^{10} kJ/g x 2 x 0.000549 g = -9.88 x 10^7 kJ

b. (9.88 x 10^7 kJ)/(6.022 x 10^{23}) = 1.64 x 10^{-16} kJ

51. Decay is too slow; $t_{\frac{1}{2}} = 0.693/(1.51 \times 10^{-3}\ yr^{-1}) = 459$ yr

53. n $Tl^{+} = 815$ cpm $\times \dfrac{1.00\ g\ Tl}{5.53 \times 10^{5}\ cpm} \times \dfrac{1\ mol\ Tl}{204\ g} = 7.22 \times 10^{-6}$ mol Tl

$[O_2] = \dfrac{7.22 \times 10^{-6}\ mol\ Tl}{0.0250\ L} \times \dfrac{\frac{1}{2}\ mol\ O_2}{2\ mol\ Tl} = 7.22 \times 10^{-5}$ mol/L

55. $k = \dfrac{0.693}{12.3\ yr} \times \dfrac{1\ yr}{365\ d} \times \dfrac{1\ d}{24\ hr} \times \dfrac{1\ hr}{3600\ s} = 1.79 \times 10^{-9}/s$

$4.22 \times 10^{3} = 1.79 \times 10^{-9} \times$ no. tritium atoms

no. tritium atoms $= 2.36 \times 10^{12}$

total no. H atoms $= \dfrac{100.0}{18.02} \times 2 \times 6.022 \times 10^{23} = 6.684 \times 10^{24}$ atoms

% tritium $= 2.36 \times 10^{12} \times 100/(6.684 \times 10^{24}) = 3.53 \times 10^{-11}$ %

57. 5.0 mL $\times \dfrac{1.7 \times 10^{5}\ cpm}{1.3 \times 10^{3}\ cpm} = 6.5 \times 10^{2}$ mL

59. From problem 39a, ΔE per gram $= -7.71 \times 10^{7}$ kJ

$C_8H_{18}(1) + 25/2\ O_2(g) \longrightarrow 8CO_2(g) + 9H_2O(g)$

$\Delta H = 9(-241.8\ kJ) + 8(-393.5\ kJ) + 249.9\ kJ = -5074.3$ kJ

7.71×10^{7} kJ $\times \dfrac{1\ mol\ C_8H_{18}}{5074.3\ kJ} \times \dfrac{114.2\ g\ C_8H_{18}}{1\ mol\ C_8H_{18}} \times \dfrac{1\ L}{703\ g} = 2.47 \times 10^{3}$ L

61. $\ln \dfrac{25.00}{x} = \dfrac{0.693 \times 75.0\ hr}{138\ d} \times \dfrac{1\ d}{24\ hr} = 0.0157$

Solving, $x = 24.61$ g; $\Delta x = 0.39$ g

0.39 g Po-210 $\times \dfrac{1\ mol\ Po\text{-}210}{210\ g} \times \dfrac{1\ mol\ He}{1\ mol\ Po} = 1.86 \times 10^{-3}$ mol He

$V = \dfrac{(1.86 \times 10^{-3}\ mol)(0.0821\ L\cdot atm/mol\cdot K)(298\ K)}{1.20\ atm} = 3.8 \times 10^{-2}$ L

63. $\dfrac{20 \times 10^{-12}\ Ci}{1\ L} \times \dfrac{3.700 \times 10^{10}\ atom/s}{1\ Ci} = 0.74$ atom/s

63. (cont.)

$$t_{\frac{1}{2}} = 3.82 \text{ d} \times \frac{24 \text{ hr}}{1 \text{ d}} \times \frac{3600 \text{ s}}{1 \text{ hr}} = 3.30 \times 10^5 \text{ s}$$

$$0.74 \text{ atom/s} = 0.693 \times n_t/(3.30 \times 10^5 \text{ s}); \quad n_t = 3.52 \times 10^5 \text{ atoms}$$

$$[Rn] = 3.52 \times 10^5 \frac{\text{atom}}{L} \times \frac{1 \text{ mol}}{6.022 \times 10^{23} \text{ atoms}} = 5.8 \times 10^{-19} \text{ mol/L}$$

64. a. $\text{activity} = \frac{5.5 \times 10^{-11}}{\text{min}} \times 1.00 \text{ g} = 5.5 \times 10^{-11} \text{ g/min}$

$\text{mass} = 5.5 \times 10^{-11} \text{ g/min} \times 25 \text{ min} = 1.4 \times 10^{-9} \text{ g}$

b. Δ m per mole = 234.9934 g + 4.0015 g − 239.0006 g = −0.0057 g

$\Delta E = -(9.00 \times 10^{10} \text{kJ/g})(0.0057/239)(1.4 \times 10^{-9} \text{ g}) = -3.0 \times 10^{-3} \text{ kJ}$

c. $3.0 \times 10^{-3} \text{ kJ} \times \frac{10^3 \text{ J}}{1 \text{ kJ}} \times \frac{1 \text{ rad}}{10^{-2} \text{ J/kg}} \times \frac{1}{75 \text{ kg}} \times \frac{10 \text{ rem}}{1 \text{ rad}} = 40 \text{ rem}$

65. a. $E = \dfrac{(8.99 \times 10^9)(1.60 \times 10^{-19})^2}{2 \times 10^{-15}} = 1.2 \times 10^{-13} \text{ J}$

b. $\dfrac{1.2 \times 10^{-13}}{2} = \dfrac{v^2}{2}(2.014 \times 10^{-3})/(6.022 \times 10^{23})$

Solving, $v = 6 \times 10^6 \text{ m/s}$

66. a. Δ m = 4.0015 g − 2(2.01355 g) = −0.02560 g

$\Delta E = 9.00 \times 10^{10} \times \dfrac{-0.02560}{4.027} = -5.72 \times 10^8 \text{ kJ/g}$

b. $\Delta E = 5.72 \times 10^8 \text{ kJ/g} \times 1.3 \times 10^{24} \text{ g} \times 1.7 \times 10^{-5} = 1.3 \times 10^{28} \text{ kJ}$

c. $(2.3 \times 10^{17})/(1.3 \times 10^{28}) = 1.8 \times 10^{-11}$

CHAPTER 20
Chemistry of the Metals

LECTURE NOTES

This is a descriptive chapter which concentrates upon the chemistry of the Group 1 and Group 2 metals on the one hand and the transition metals on the other. (Recall that the complex ions of the transition metals were covered in Chapter 15). The chapter ends with alloys, introduced in hopes of raising student interest.

This chapter is readily covered in $1\frac{1}{2}$ - 2 lectures.

LECTURE 1

I **Metallurgy**

A. **Chloride ores**; convert to metal by electrolysis

$$NaCl(1) \longrightarrow Na(1) + \tfrac{1}{2}Cl_2(g)$$

Volume of chlorine at 25°C, 1.00 atm produced by current of 25.0 A in one hour?

$$n\ Cl_2 = (25.0)(3600) \times \frac{1\ mol\ e^-}{9.648 \times 10^4\ C} \times \frac{\tfrac{1}{2}\ mol\ Cl_2}{1\ mol\ e^-} = 0.466\ mol$$

V (from ideal gas law) = 11.4 L

B. **Oxide ores**; most often reduce with CO

Metallurgy of iron:

$$Fe_2O_3(s) + 3CO(g) \longrightarrow 2Fe(s) + 3CO_2(g)$$

$$\Delta H° = -26.8\ kJ \qquad \Delta S° = +11.5\ J/K$$

Spontaneous at all temperatures; carried out at high T to make

198

Saunders College Publishing

reaction go quickly. Limestone added to make slag of $CaSiO_3$ with SiO_2 impurity.

   c. <u>Sulfide ores</u>; roast in air to form metal or oxide

$$HgS(s) + O_2(g) \longrightarrow Hg(g) + SO_2(g)$$

$$2ZnS(s) + 3\,O_2(g) \longrightarrow 2ZnO(s) + 2SO_2(g)$$

Metallurgy of copper; copper(I) sulfide concentrated by flotation, roasted to copper, purified by electrolysis.

II <u>Reactions of Group 1, Group 2 Metals</u>

   A. <u>Reaction with hydrogen</u>; form metal hydrides (H^- ion)

$$2Na(s) + H_2(g) \longrightarrow 2NaH(s)$$

$$Ca(s) + H_2(g) \longrightarrow CaH_2(s)$$

   B. <u>Reaction with water</u>; evolve H_2, form OH^- ions

$$2Na(s) + 2H_2O(l) \longrightarrow H_2(g) + 2Na^+(aq) + 2\,OH^-(aq)$$

$$Ca(s) + 2H_2O(l) \longrightarrow H_2(g) + Ca^{2+}(aq) + 2\,OH^-(aq)$$

   c. <u>Reaction with oxygen</u>

1. Lithium and Group 2 metals yield normal oxides (O^{2-} ion)

2. Na, Ba yield peroxides (O_2^{2-} ion)

3. K, Rb, Cs yield superoxides (O_2^- ion)

Write balanced equations for reaction of O_2 with Mg, Na, K

$$2\,Mg(s) + O_2(g) \longrightarrow 2MgO(s)$$

$$2Na(s) + O_2(g) \longrightarrow Na_2O_2(s)$$

$$K(s) + O_2(g) \longrightarrow KO_2(s)$$

<u>LECTURE 2</u>

I <u>Redox Chemistry of the Transition Metals</u>

   A. <u>Reaction with acid</u>

199

1. $Zn(s) + 2H^+(aq) \longrightarrow Zn^{2+}(aq) + H_2(g)$

   occurs if $E^\circ_{ox} > 0$

2. $3Cu(s) + 2NO_3^-(aq) + 8H^+(aq) \longrightarrow 3Cu^{2+}(aq) + 2NO(g) + 4H_2O$

   occurs if $E^\circ_{ox} > -0.964$ V

B. Cations may be unstable in water because of

   1. Reaction with water

$$Co^{3+}(aq) + e^- \longrightarrow Co^{2+}(aq) \qquad E^\circ_{red} = +1.953 \text{ V}$$

$$\frac{H_2O \longrightarrow \tfrac{1}{2}O_2(g) + 2H^+(aq) + 2e^-}{2Co^{3+}(aq) + H_2O \longrightarrow 2Co^{2+}(aq) + \tfrac{1}{2}O_2(g) + 2H^+(aq)} \qquad E^\circ_{ox} = -1.229 \text{ V}$$

$$E^\circ = +0.724 \text{ V}$$

   2. Disproportionation

$$Cu^+(aq) + e^- \longrightarrow Cu(s) \qquad\qquad +0.518 \text{ V}$$

$$\frac{Cu^+(aq) \longrightarrow Cu^{2+}(aq) + e^-}{2Cu^+(aq) \longrightarrow Cu^{2+}(aq) + Cu(s)} \qquad\qquad \begin{array}{c} -0.161 \text{ V} \\ +0.357 \text{ V} \end{array}$$

   3. Oxidation by dissolved oxygen

$$Fe^{2+}(aq) \longrightarrow Fe^{3+}(aq) + e^- \qquad\qquad -0.769 \text{ V}$$

$$\frac{\tfrac{1}{2}O_2(g) + 2H^+(aq) + 2e^- \longrightarrow H_2O}{2Fe^{2+}(aq) + \tfrac{1}{2}O_2(g) + 2H^+(aq) \longrightarrow 2Fe^{3+}(aq) + H_2O} \qquad \begin{array}{c} +1.229 \text{ V} \\ +0.460 \text{ V} \end{array}$$

C. <u>Oxyanions</u> (CrO_4^{2-}, $Cr_2O_7^{2-}$, MnO_4^-)

   1. $2CrO_4^{2-}(aq) + 2H^+(aq) \rightleftharpoons Cr_2O_7^{2-}(aq) + H_2O$

      yellow   red

      Dichromate ion is stable in acid, chromate stable in basic or neutral solution.

   2. Both dichromate and permanganate ions are powerful oxidizing agents in acidic solution.

$$Cr_2O_7^{2-}(aq) + 14H^+(aq) + 6e^- \longrightarrow 2Cr^{3+}(aq) + 7H_2O$$

$$E_{red} = 1.33 \text{ V} - \frac{0.0591}{6} \log_{10} \frac{[Cr^{3+}]^2}{[Cr_2O_7^{2-}] \times [H^+]^{14}}$$

200

If $[Cr^{3+}] = [Cr_2O_7^{2-}] = 1.0$ M, then:

$$E_{red} = 1.33 \text{ V} - 0.14 \text{ pH}$$

E_{red} decreases from 1.33 V in 1 M H^+ to +0.35 V at pH 7

DEMONSTRATIONS

1. Preparation of Li metal: GILB M 22
2. Copper to silver to gold: SHAK 4 263
3. Alkali metals: GILB A 4
4. Reaction of sodium with water: GILB A 31; J. Chem. Educ. 69 418 (1992)
5. Reaction of magnesium with steam: GILB M 54
6. Reaction of magnesium with carbon dioxide: GILB M 55
7. Oxidation states of Fe, Sn and Hg: GILB G 20
8. Cu(II) - Cu(I) equilibrium: GILB J 26
9. Chromate - dichromate equilibrium: GILB J 25
10. Decomposition of ammonium dichromate: GILB A 7, H 13; SHAK 1 81
11. Oxidation states of Mn: GILB G 19, M 265
12. Properties of Wood's metal: GILB M 176

<u>VIDEO</u>

1. Copper, silver, and gold: SHAK 11
2. Reaction of potassium with bromine: Falcon 17
3. Reaction of magnesium with fluorine: FALCON 18
4. Oxidation and reduction of copper: JCES 9
5. Ammonium dichromate volcano: JCES 1

PROBLEMS

1. $2NaCl(s) \longrightarrow 2Na(l) + Cl_2(g)$

$$n\ Cl_2 = 1.00\ g\ Na \times \frac{1\ mol\ Na}{22.99\ g\ Na} \times \frac{1\ mol\ Cl_2}{2\ mol\ Na} = 0.0217\ mol\ Cl_2$$

V (from ideal gas law) = 0.486 L

3. a. $2NiS(s) + 3\ O_2(g) \longrightarrow 2NiO(s) + 2SO_2(g)$

b. $NiO(s) + CO(g) \longrightarrow Ni(s) + CO_2(g)$

5. $\Delta G^\circ = \Delta G_f^\circ\ CO_2 - \Delta G_f^\circ\ CO - \Delta G_f^\circ\ NiO = -45.5$ kJ; yes

7. a. $4Au(s) + 8CN^-(aq) + O_2(g) + 2H_2O \longrightarrow 4Au(CN)_2^-(aq) + 4\ OH^-(aq)$

b. $Zn(s) + 2Au(CN)_2^-(aq) \longrightarrow Zn(CN)_4^{2-}(aq) + 2Au(s)$

9. $$n\ O_2 = 10^6\ g\ Fe_2O_3 \times 0.92 \times \frac{1\ mol\ Fe_2O_3}{159.70\ g\ Fe_2O_3} \times \frac{3\ mol\ CO}{1\ mol\ Fe_2O_3}$$

$$\times \frac{\tfrac{1}{2}\ mol\ O_2}{1\ mol\ CO} = 8.64 \times 10^3\ mol\ O_2$$

$V\ O_2(g)$ from ideal gas law = 2.11×10^5 L

$$V\ air = 2.11 \times 10^5\ L\ O_2 \times \frac{1\ L\ air}{0.21\ L\ O_2} \times \frac{1\ ft^3}{28.32\ L} = 3.6 \times 10^4\ ft^3$$

11. $MS(s) + 3/2\ O_2(g) \longrightarrow MO(s) + SO_2(g)$

Let $\mathcal{M}$ = molar mass of M

$$\frac{\mathcal{M} + 16.00}{\mathcal{M} + 32.07} = \frac{2.368}{2.876} = 0.8234;\ \mathcal{M} = 58.95\ g/mol$$

13. a. Sr_3N_2; strontium nitride

b. $SrBr_2$; strontium bromide

c. $Sr(OH)_2$; strontium hydroxide

d. SrO; strontium oxide

15. a. $Mg(s) + Cl_2(g) \longrightarrow MgCl_2(s)$; magnesium chloride

b. $BaO_2(s) + 2H_2O \longrightarrow H_2O_2(aq) + Ba^{2+}(aq) + 2\ OH^-(aq)$
hydrogen peroxide, barium hydroxide

c. $2Li(s) + S(s) \longrightarrow Li_2S(s)$; lithium sulfide

d. $2Na(s) + 2H_2O \longrightarrow H_2(g) + 2Na^+(aq) + 2\ OH^-(aq)$
    hydrogen, sodium hydroxide

17. $CaH_2(s) + 2H_2O \longrightarrow 2H_2(g) + Ca^{2+}(aq) + 2\ OH^-(aq)$

$$n\ H_2 = \frac{(1.10\ atm)(25.0\ L)}{(0.0821\ L \cdot atm/mol \cdot K)(298\ K)} = 1.12\ mol$$

$$mass = 1.12\ mol\ H_2 \times \frac{1\ mol\ CaH_2}{2\ mol\ H_2} \times \frac{42.10\ g\ CaH_2}{1\ mol\ CaH_2} = 23.6\ g\ CaH_2$$

19. a. $2CrO_4^{2-}(aq) + 2H^+(aq) \longrightarrow Cr_2O_7^{2-}(aq) + H_2O$

    b. $4MnO_4^-(aq) + 2H_2O \longrightarrow 4MnO_2(s) + 4\ OH^-(aq) + 3\ O_2(g)$

21. $Hg(1) + 4Cl^-(aq) + 2NO_3^-(aq) + 4H^+(aq) \longrightarrow HgCl_4^{2-}(aq) + 2NO_2(g)$
$$+\ 2H_2O$$

23. a. $Cu(s) + 2NO_3^-(aq) + 4H^+(aq) \longrightarrow Cu^{2+}(aq) + 2NO_2(g) + 2H_2O$

    b. $2Cr(OH)_3(s) + 3ClO^-(aq) + 4\ OH^-(aq) \longrightarrow 2CrO_4^{2-}(aq) + 3Cl^-(aq)$
$$+\ 5H_2O$$

25. a. $E^\circ = +0.402$ V; spontaneous

    b. $E^\circ = +0.912$ V; spontaneous

    c. $E^\circ = +0.282$ V; spontaneous

    d. $E^\circ = -0.799$ V; nonspontaneous

    e. $E^\circ = -1.498$ V; nonspontaneous

27. Au^+: $E^\circ = 1.695\ V - 1.400\ V = +0.295\ V$

29. a. $3Fe^{2+}(aq) \longrightarrow 2Fe^{3+}(aq) + Fe(s)$; $E^\circ = -1.178$ V

    $\ln K = -2(1.178)/0.0257$; $K = 2 \times 10^{-40}$

    b. $\dfrac{[Fe^{3+}]^2}{[Fe^{2+}]^3} = 2 \times 10^{-40}$; $[Fe^{3+}]^2 = 2 \times 10^{-43}$; $[Fe^{3+}] = 4 \times 10^{-22}$ M

31. a. $n\ Cu = 75.0\ g/(63.55\ g/mol) = 1.18\ mol$

    $n\ Ni = 25.0\ g/(58.69\ g/mol) = 0.426\ mol$

    $X\ Cu = \dfrac{1.18}{1.18 + 0.426} = 0.735$

b. 0.925×1.00 g Ag $\times \dfrac{1 \text{ mol Ag}}{107.9 \text{ g Ag}} \times \dfrac{6.022 \times 10^{23} \text{ atoms}}{1 \text{ mol Ag}}$

$= 5.16 \times 10^{21}$ atoms

33. $2Na(s) + 2H_2O \longrightarrow H_2(g) + 2Na^+(aq) + 2\ OH^-(aq)$

$n\ H_2 = \dfrac{(2.73 \text{ L})(752/760 \text{ atm})}{(0.0821 \text{ L·atm/mol·K})(295 \text{ K})} = 0.112$ mol

mass Na $= 0.112$ mol $H_2 \times \dfrac{2 \text{ mol Na}}{1 \text{ mol } H_2} \times \dfrac{22.99 \text{ g Na}}{1 \text{ mol Na}} = 5.13$ g Na

35. $[Pb^{2+}] = (1.7 \times 10^{-5})/(2.0 \times 10^{-1})^2 = 4.2 \times 10^{-4}$ M

37. $Fe_2O_3(s) + 3H_2(g) \longrightarrow 2Fe(s) + 3H_2O(g)$

$\Delta H° = 3(-241.8 \text{ kJ}) + 824.2 \text{ kJ} = +98.8$ kJ

$\Delta S° = +0.1415$ kJ/K

$T = 98.8/0.1415 = 698$ K $= 425°C$

39. 2.2×10^5 g Ni $\times \dfrac{90.76 \text{ g NiS}}{58.69 \text{ g Ni}} = 3.4 \times 10^5$ g NiS

41. a. $Fe^{2+}(aq) \longrightarrow Fe^{3+}(aq) + e^-$

$MnO_4^-(aq) + 5e^- + 8H^+(aq) \longrightarrow Mn^{2+}(aq) + 4H_2O$

$5Fe^{2+}(aq) + MnO_4^-(aq) + 8H^+(aq) \longrightarrow 5Fe^{3+}(aq) + Mn^{2+}(aq) + 4H_2O$

b. $E° = +0.743$ V

c. mass Fe $= (0.05563 \times 0.0200)$mol $MnO_4^- \times \dfrac{5 \text{ mol Fe}}{1 \text{ mol } MnO_4^-} \times \dfrac{55.85 \text{ g Fe}}{1 \text{ mol Fe}}$

$= 0.311$ g

% Fe $= \dfrac{0.311}{0.3500} \times 100 = 88.8\%$

43. Let x = mass BaO; $22.38 - x$ = mass BaO_2

$20.00 \text{ g} = \dfrac{137.3}{153.3}x + \dfrac{137.3}{169.3}(22.38 \text{ g} - x)$

$x = 21.86$ g; 98% BaO, 2% BaO_2

Saunders College Publishing

44. a. $Fe(OH)_3(s) + 3H_2C_2O_4(aq) \longrightarrow Fe(C_2O_4)_3^{3-}(aq) + 3H^+(aq) + 3H_2O$

b. $1.0 \text{ g } Fe(OH)_3 \times \dfrac{1 \text{ mol } Fe(OH)_3}{106.88 \text{ g } Fe(OH)_3} \times \dfrac{3 \text{ mol } H_2C_2O_4}{1 \text{ mol } Fe(OH)_3} = 0.0281 \text{ mol } H_2C_2O_4$

$V = 0.0281 \text{ mol}/(0.10 \text{ mol/L}) = 2.8 \times 10^2 \text{ mL}$

45. $6Fe^{2+}(aq) + Cr_2O_7^{2-}(aq) + 14H^+(aq) \longrightarrow 6Fe^{3+}(aq) + 2Cr^{3+}(aq) + 7H_2O$

$n \text{ } Fe^{2+} = (0.01350 \times 0.100) \text{ mol } Cr_2O_7^{2-} \times \dfrac{6 \text{ mol } Fe^{2+}}{1 \text{ mol } Cr_2O_7^{2-}}$

$= 8.10 \times 10^{-3} \text{ mol } Fe^{2+}$

$n \text{ } Fe^{2+} \text{ used} = 0.07500 \times 0.125 - 0.00810 = 0.001275$

$5Fe^{2+}(aq) + MnO_4^-(aq) + 8H^+(aq) \longrightarrow 5Fe^{3+}(aq) + Mn^{2+}(aq) + 4H_2O$

$n \text{ } MnO_4^- = 1.275 \times 10^{-3} \text{ mol } Fe^{2+} \times \dfrac{1 \text{ mol } MnO_4^-}{5 \text{ mol } Fe^{2+}} = 2.55 \times 10^{-4} \text{ mol}$

$\text{mass } Mn = 2.55 \times 10^{-4} \text{ mol} \times \dfrac{54.94 \text{ g}}{1 \text{ mol}} = 0.0140 \text{ g}$

$\% \text{ } Mn = \dfrac{0.0140}{0.500} \times 100 = 2.80\%$

46. $\Delta H° = +520.0 \text{ kJ}; \quad \Delta S° = +0.1840 \text{ kJ/K}$

$T = 2830 \text{ K} \approx 2560°C$

47. $Cr_2O_7^{2-}(aq) + 2 \text{ } OH^-(aq) \longrightarrow 2CrO_4^{2-}(aq) + H_2O$

$2Ag^+(aq) + CrO_4^{2-}(aq) \longrightarrow Ag_2CrO_4(s)$

$Ag_2CrO_4(s) + 4NH_3(aq) \longrightarrow 2Ag(NH_3)_2^+(aq) + CrO_4^{2-}(aq)$

$2Ag(NH_3)_2^+(aq) + CrO_4^{2-}(aq) + 4H^+(aq) \longrightarrow Ag_2CrO_4(s) + 4NH_4^+(aq)$

CHAPTER 21
Chemistry of the Nonmetals

LECTURE NOTES

This is a heavily descriptive chapter that puts a great deal of emphasis on writing equations, particularly net ionic equations. Students probably have more trouble with that topic than with any other in general chemistry. It is certainly true that many of the equations in this chapter have appeared before, albeit in a different context. Remember, though, most of the students had trouble with these equations the first time around, so a little review won't hurt.

As in every descriptive chapter, not every topic needs to be covered in lecture. With that in mind. two lectures should be sufficient for this chapter.

<u>LECTURE 1</u>

I <u>Preparation of the Elements</u>

   A. Nitrogen and oxygen from liquid air.

   B. Sulfur from underground deposits; Frasch process

   C. Halogens

      Fluorine by electrolysis of HF, chlorine by electrolysis of aqueous sodium chloride (Chapter 18). Bromine and iodine prepared using chlorine:

$$Cl_2(g) + 2Br^-(aq) \longrightarrow Br_2(l) + 2Cl^-(aq); \quad E^\circ = +0.283 \text{ V}$$

$$Cl_2(g) + 2I^-(aq) \longrightarrow I_2(s) + 2Cl^-(aq); \quad E^\circ = +0.826 \text{ V}$$

II <u>Allotropy</u>

   A. O_2 vs O_3. Electronic structures

   B. White and red phosphorus; structure and properties

206

C. Sulfur   Rhombic and monoclinic structures.   Behavior of liquid
sulfur; polymerization between 160-250°C.   Chains:

$$\cdot S \diagup \overset{S}{\diagdown} \underset{S}{\diagup} \overset{S}{\diagdown} \underset{S}{\diagup} \overset{S}{\diagdown} \underset{S}{\diagup} S \cdot$$

III Hydrogen Compounds

A. Ammonia

Bronsted base: $NH_3 + H_2O \rightleftharpoons NH_4^+(aq) + OH^-(aq)$

Lewis base: $2NH_3(aq) + Ag^+(aq) \longrightarrow Ag(NH_3)_2^+(aq)$

Precipitating agent:

$$Fe^{3+}(aq) + 3NH_3(aq) + 3H_2O \longrightarrow Fe(OH)_3(s) + 3NH_4^+(aq)$$

Can also act as reducing agent, but not as oxidizing agent

B. Hydrogen sulfide

Bronsted acid, precipitating agent, reducing agent

C. Hydrogen peroxide

$$H_2O_2(aq) + 2H^+(aq) + 2e^- \longrightarrow 2H_2O; \quad E^o_{red} = +1.763 \text{ V}$$

$$H_2O_2(aq) \longrightarrow O_2(g) + 2H^+(aq) + 2e^-; \quad E^o_{ox} = -0.695 \text{ V}$$

strong oxidizing agent, weak reducing agent

D. HF, HCl

Reaction with carbonate ion:

$$2H^+(aq) + CO_3^{2-}(aq) \longrightarrow CO_2(g) + H_2O$$

$$2HF(aq) + CO_3^{2-}(aq) \longrightarrow CO_2(g) + H_2O + 2F^-(aq)$$

LECTURE 2

I Oxygen Compounds

A. Molecular structures of oxides of N, P, S

B. Reaction with water (acid anhydrides)

$$N_2O_5(g) + H_2O(l) \longrightarrow 2HNO_3(l); \text{ oxidation state unchanged}$$

207

$$CO_2(g) + H_2O \longrightarrow H_2CO_3(aq)$$

II. <u>Oxoacids, Oxoanions</u>

A. Acid strength

   Increases with electronegativity and oxidation number of central atom

$$HClO_4 \;>\; HClO_3 \;>\; HClO_2 \;>\; HClO$$
$$\text{very strong} \quad \text{strong} \qquad \text{weak} \qquad \text{very weak}$$

$$HClO \;>\; HBrO \;>\; HIO$$

   Rationale: strength depends upon how readily O-H bond is broken to
   form proton. Increase in electronegativity or oxidation number tends
   to draw electrons away from O atom bonded to hydrogen, making ionization easier.

B. Oxidizing and reducing strength

   1. Species in highest oxidation state (nitrate, sulfate) can act only
      as oxidizing agent. In intermediate state (nitrite, sulfite), can
      act as either oxidizing or reducing agent.

   $$NO_2^-(aq) + H_2O \longrightarrow NO_3^-(aq) + 2H^+(aq) + 2e^-$$

   $$NO_2^-(aq) + 2H^+(aq) + e^- \longrightarrow NO(g) + H_2O$$

   2. Oxidizing strength increases with H^+ ion concentration

   $$NO_3^-(aq) + 4H^+(aq) + 3e^- \longrightarrow NO(g) + H_2O$$

$[H^+]$	1.0 M	1.0×10^{-7} M	1.0×10^{-14} M
E_{red}	+0.964 V	+0.412 V	−0.139 V

C. Nitric acid Strong acid, strong oxidizing agent. Can be reduced to
   NO_2, NO, or even NH_4^+.

D. Sulfuric acid Strong acid, dehydrating agent (effect on sugar).
   relatively weak oxidizing agent (E°_{red} = +0.155 V).

E. Phosphoric acid

	H_3PO_4	$H_2PO_4^-$	HPO_4^{2-}	PO_4^{3-}
K_a	7×10^{-3}	6×10^{-8}	5×10^{-13}	
K_b		1×10^{-12}	2×10^{-7}	2.2×10^{-2}

DEMONSTRATIONS

1. Allotropic forms of sulfur: GILB B 7, M 206; SHAK 1 243
2. Properties of ozone: GILB D 27
3. Conversion of red to white phosphorus: GILB M 170
4. Phosphorus and oxygen: GILB M 158; SHAK 1 74, 186
5. Reaction of white phosphorus with copper(II): GILB G 13
6. Preparation of chlorine: SHAK 2 220
7. Reaction of chlorine with elements: GILB M 232
8. Redox reactions of halogens: GILB G 21, M 229
9. Bleaching with chlorine: J. Chem. Educ. 70 154 (1993)
10. Preparation and properties of ammonia: GILB M 138; SHAK 2 202
11. Etching glass with HF: GILB M 115; SHAK 3 80
12. Hydrogen peroxide rocket: GILB I 234
13. Reactions of hydrogen peroxide: J. Chem. Educ. 71 433 (1994)
14. Nitrogen triiodide: GILB M 137; SHAK 1 96; J. Chem. Educ. 70 943 (1993)
15. Oxidizing strength of nitric acid: GILB G 10
16. Properties of nitric and sulfuric acids: SHAK 3 70
17. Reaction of sulfuric acid with sugar: SHAK 1 77

<u>VIDEO</u>

1. Plastic sulfur: JCES 2
2. Writing with white phosphorus: JCES 10
3. Burning phosphorus: Falcon 30
4. Halides and halogens: Falcon 12
5. Etching glass: Falcon 14
6. Nitrogen triiodide: SHAK 27; JCES 20
7. Reaction of sulfuric acid with sugar: SHAK 26

PROBLEMS

1. a. periodic acid   b. bromite ion   c. hypoiodous acid

   d. sodium chlorate

3. a. $HClO_3$ b. HIO_4 c. $HBrO$ d. HI

5. a. NH_3 b. H_2S c. $NaCl$

7. a. N_2O_5 b. N_2O_3 c. SO_3

9. a. NH_3 b. N_2O c. H_2O_2 d. SO_3

11. a. NH_3 b. P_2H_4 c. H_2O_2

13. a. S^{2-} b. HSO_3^-, SO_3^{2-} c. H_2SO_3, H_2SO_4, H_2S

15. a. $2HF(l) \longrightarrow H_2(g) + F_2(g)$

   b. $H_2O_2(aq) + 2I^-(aq) + 2H^+(aq) \longrightarrow 2H_2O + I_2(s)$

17. a. $3I_2(s) + 6\ OH^-(aq) \longrightarrow IO_3^-(aq) + 5I^-(aq) + 3H_2O$

   b. $4Cl_2(g) + 8\ OH^-(aq) \longrightarrow 7Cl^-(aq) + ClO_4^-(aq) + 4H_2O$

19. a. $Cl_2(g) + 2I^-(aq) \longrightarrow 2Cl^-(aq) + I_2(s)$

   b. $F_2(g) + 2Br^-(aq) \longrightarrow 2F^-(aq) + Br_2(l)$

   c. N. R.

   d. $Br_2(l) + 2I^-(aq) \longrightarrow 2Br^-(aq) + I_2(s)$

21. a. $2HF(l) \longrightarrow H_2(g) + F_2(g)$

   b. $Cl_2(g) + 2Br^-(aq) \longrightarrow 2Cl^-(aq) + Br_2(l)$

   c. $NH_3(aq) + H^+(aq) \longrightarrow NH_4^+(aq)$

23. a. $Cu^{2+}(aq) + 4NH_3(aq) \longrightarrow Cu(NH_3)_4^{2+}(aq)$

   b. $H^+(aq) + NH_3(aq) \longrightarrow NH_4^+(aq)$

c. $Al^{3+}(aq) + 3NH_3(aq) + 3H_2O \longrightarrow Al(OH)_3(s) + 3NH_4^+(aq)$

25. a. $H^+(aq) + OH^-(aq) \longrightarrow H_2O$

b. $Ag(s) + NO_3^-(aq) + 2H^+(aq) \longrightarrow Ag^+(aq) + NO_2(g) + H_2O$

c. $5Cd(s) + 2NO_3^-(aq) + 12H^+(aq) \longrightarrow 5Cd^{2+}(aq) + N_2(g) + 6H_2O$

27. b, c

29. a. O_3 vs O_2 molecules

b. P_4 molecules vs network covalent

c. S_8 molecules vs long chains

31. a. (Lewis structure: N double-bonded to one O, single-bonded to another O)   b. $\cdot N = O$ c. $O - S = O$ d. $O - S - O$ with O below

33. a, b, c

35. a. H - O - N - O with O below (nitric acid)   b. H - O - S - O - H with O above and O below (sulfuric acid)   c. H - O - P - O - H with O above and O - H below (phosphoric acid)

37. a. H - O - Br - O with O above and O below (perbromic acid)   b. H - N - N - H with H below each N (hydrazine)   c. H - O - P - O - H with O above and O - H below

39. a. 1.00×10^3 g Br^- $\times \dfrac{1 \text{ g seawater}}{65 \times 10^{-6} \text{ g } Br^-} \times \dfrac{1 \text{ lb}}{453.6 \text{ g}} \times \dfrac{1 \text{ ft}^3}{64.0 \text{ lb}}$

$= 5.3 \times 10^2 \text{ ft}^3$

b. $n\, Cl_2 = 1.00 \times 10^3$ g $Br^- \times \dfrac{1 \text{ mol } Br^-}{79.90 \text{ g } Br^-} \times \dfrac{1 \text{ mol } Cl_2}{2 \text{ mol } Br^-} = 6.26 \text{ mol } Cl_2$

V (ideal gas law) $= 1.50 \times 10^2$ L

41. mass I_2 = 0.200 g MnO_2 x $\dfrac{1 \text{ mol } MnO_2}{86.94 \text{ g } MnO_2}$ x $\dfrac{1 \text{ mol } I_2}{1 \text{ mol } MnO_2}$ x $\dfrac{253.8 \text{ g } I_2}{1 \text{ mol } I_2}$

$\qquad$ = 0.584 g I_2

43. $NH_4NO_3(s) \longrightarrow N_2(g) + 2H_2O(g) + \frac{1}{2} O_2(g)$

$\qquad$ n gas = 1.000 x 10^3 g NH_4NO_3 x $\dfrac{1 \text{ mol } NH_4NO_3}{80.05 \text{ g } NH_4NO_3}$ x $\dfrac{3.5 \text{ mol gas}}{1 \text{ mol } NH_4NO_3}$

$\qquad$ = 43.72 mol

$\qquad$ P = 2.77 x 10^3 atm

45. $2NaNO_3(s) \longrightarrow 2NaNO_2(s) + O_2(g)$

$\qquad$ n O_2 = $\dfrac{(0.1250 \text{ L})(731/760 \text{ atm})}{(0.0821 \text{ L·atm/mol·K})(296 \text{ K})}$ = 4.95 x 10^{-3} mol

$\qquad$ mass $NaNO_3$ = 4.95 x 10^{-3} mol O_2 x $\dfrac{2 \text{ mol } NaNO_3}{1 \text{ mol } O_2}$ x $\dfrac{85.00 \text{ g } NaNO_3}{1 \text{ mol } NaNO_3}$

$\qquad$ = 0.841 g $NaNO_3$

% $NaNO_3$ = $\dfrac{0.841}{1.500}$ x 100% = 56.1%

47. $[H^+]$ x $[Br^-]$ x $[HBrO]$ = 1.2 x 10^{-9}

$\qquad$ $[H^+]$ = (1.2 x $10^{-9})^{1/3}$ = 1.1 x 10^{-3} M ; pH = 2.97

49. K = $(0.7324)^2/(2.80 \times 10^{-3})^2$ = 6.84 x 10^4

51. $[Ba^{2+}]$ x $[F^-]^2$ = 1.8 x 10^{-7}

$\qquad$ $[F^-]^3$ = 3.6 x 10^{-7}; $[F^-]$ = 7.1 x 10^{-3} M

53. $\Delta H°$ = -795.4 kJ + 1002.8 kJ = +207.4 kJ

$\qquad$ $\Delta S°$ = +0.1161 kJ/K + 0.2862 kJ/K - 0.2230 kJ/K - 0.3030 kJ/K

$\qquad$ = -0.1237 kJ/K

$\qquad$ $\Delta G°$ = 207.4 kJ + 298 K(0.1237 kJ/K) = +244.3 kJ; nonspontaneous

55. $\Delta H°$ = -332.6 kJ + 320.1 kJ = -12.5 kJ

$\qquad$ $\Delta G°$ = -8.31 x 10^{-3} kJ/K x 298 K x ln(6.9 x 10^{-4}) = +18.0 kJ

55. (cont.)

$$\Delta S° = \frac{-30.5 \text{ kJ}}{298 \text{ K}} = -0.0138 \text{ kJ/K} - S° \text{ HF(aq)}$$

$$S° \text{ HF(aq)} = 0.088 \text{ kJ/K}$$

57. a. $\Delta H° = 4(90.2 \text{ kJ}) + 6(-241.8 \text{ kJ}) - 4(-46.1 \text{ kJ}) = -905.6 \text{ kJ}$

    exothermic

b. positive

$$\Delta S° = 6(0.1887 \text{ kJ/K}) + 4(+0.2107 \text{ kJ/K}) - 5(+0.2050 \text{ kJ/K})$$

$$-4(0.1923 \text{ kJ/K}) = +0.1808 \text{ kJ/K}$$

c. $\Delta G° = -905.6 \text{ kJ} - 298(0.1808 \text{ kJ/K}) = -959.5 \text{ kJ}$; yes

d. none; $\Delta H°$ and $\Delta S°$ have opposite signs

59. $2I^-(aq) \longrightarrow I_2(s) + 2e^-$

$$Q = 1 \text{ mol } I_2 \times \frac{2 \text{ mol } e^-}{1 \text{ mol } I_2} \times \frac{9.648 \times 10^4 \text{ C}}{1 \text{ mol } e^-} = 1.930 \times 10^5 \text{ C}$$

$$E = 1.930 \times 10^5 \text{ C} \times 5.00 \text{ V} \times \frac{1 \text{ kJ}}{10^3 \text{ J}} = 965 \text{ kJ}$$

61. $Cl^-(aq) + H_2O \longrightarrow ClO^-(aq) + 2H^+(aq) + 2e^-$

$$n \text{ ClO}^- = 1.500 \times 10^6 \text{ g} \times 0.0500 \times \frac{1 \text{ mol NaClO}}{74.44 \text{ g NaClO}} = 1008 \text{ mol ClO}^-$$

$$n \text{ } e^- = 2016 \text{ mol}$$

$$t = \frac{2016 \text{ mol } e^- \times 9.648 \times 10^4 \text{ C/mol } e^-}{2.00 \times 10^3 \text{ C/s}} = 9.72 \times 10^4 \text{ s}$$

63. a. $E° = +0.63 \text{ V}$; yes   b. $E° = -1.104 \text{ V}$; no

c. $E° = -0.161 \text{ V}$; no   d. $E° = +0.382 \text{ V}$; yes

65. $2NO_3^-(aq) + 3SO_2(g) + 2H_2O \longrightarrow 2NO(g) + 3SO_4^{2-}(aq) + 4H^+(aq)$

$$E = +0.809 \text{ V} - \frac{0.0257}{6} \ln \frac{(0.100)^3(5.0 \times 10^{-5})^4}{(0.100)^2} = +0.988 \text{ V}$$

67. a. HClO   b. HIO_4 c. $HBrO_4$

69. $SiO_2(s) + 4HF(aq) \longrightarrow SiF_4(g) + 2H_2O$

$$n\ HF = 1.00\ g\ SiO_2 \times \frac{1\ mol\ SiO_2}{60.09\ g\ SiO_2} \times \frac{4\ mol\ HF}{1\ mol\ SiO_2} = 0.0666\ mol\ HF$$

$$V = 0.0666\ mol/(2.0\ mol/L) = 0.0333\ L = 33.3\ mL$$

71. $$\mathcal{M} = \frac{dRT}{P} = \frac{(0.8012\ g/L)(0.0821\ L \cdot atm/mol \cdot K)(973\ K)}{1.00\ atm}$$

$$= 64.0\ g/mol;\ S_2$$

73. contaminated by NO_2

75. depth of deposit, purity of sulfur, density of sulfur, molar masses of S, H_2SO_4.

76. Consider one kilogram of quartz

$$value\ of\ gold = 0.010\ g \times \frac{1\ oz}{31.1\ g} \times \frac{\$425}{1\ oz} = \$0.14$$

$$cost\ of\ HF = 1.000 \times 10^3\ g\ SiO_2 \times \frac{1\ mol\ SiO_2}{60.09\ g\ SiO_2} \times \frac{4\ mol\ HF}{1\ mol\ SiO_2}$$

$$\times \frac{20.01\ g\ HF}{1\ mol\ HF} \times \frac{1\ g\ acid}{0.50\ g\ HF} \times \frac{1\ cm^3}{1.17\ g} \times \frac{\$0.75}{1000\ cm^3} = \$1.71;\ no$$

77. $ClO^-(aq) + 2I^-(aq) + 2H^+(aq) \longrightarrow Cl^-(aq) + H_2O + I_2(s)$

$I_2(s) + 2S_2O_3^{2-}(aq) \longrightarrow 2I^-(aq) + S_4O_6^{2-}(aq)$

$$n\ I_2 = (0.02500\ L)(0.0700\ mol\ S_2O_3^{2-}/L) \times \frac{1\ mol\ I_2}{2\ mol\ S_2O_3^{2-}}$$

$$= 8.75 \times 10^{-4}\ mol\ I_2$$

$$mass\ NaClO = 8.75 \times 10^{-4}\ mol\ I_2 \times \frac{1\ mol\ ClO^-}{1\ mol\ I_2} \times \frac{74.44\ g\ NaClO}{1\ mol\ ClO^-}$$

$$= 0.0651\ g$$

$$mass\ \% = \frac{0.0651\ g}{5.00\ g} \times 100\% = 1.30\%$$

78. $NaN_3(s) \longrightarrow Na(s) + 3/2\ N_2(g)$

Assume $T = 25°C$; minimum $P = 1.00$ atm

$$n\ N_2 = \frac{(20.0\ L)(1.00\ atm)}{(0.0821\ L \cdot atm/mol \cdot K)(298\ K)} = 0.818\ mol$$

$$mass\ NaN_3 = 0.818\ mol\ N_2 \times \frac{1\ mol\ NaN_3}{1.5\ mol\ N_2} \times \frac{65.02\ g\ NaN_3}{1\ mol\ NaN_3}$$

$$= 35\ g\ NaN_3$$

CHAPTER 22
Organic Chemistry

LECTURE NOTES

This chapter is intended to give the student some idea of what organic chemistry is all about. We have tried to relate the material to topics with which students are familiar, such as the uses of acetylene and the properties of alcohol. We have deliberately avoided the encyclopedic approach, mentioning only a few functional groups with selected examples of compounds within each group. The nomenclature of alkanes is covered in some detail; alkenes, alkynes, and aromatic hydrocarbons are treated more briefly.

The material in this chapter will require between two and three lectures depending upon how much time you choose to spend on nomenclature and p.olymers (Section 22.5). If you delete the latter and don't dwell on the former, two lectures will suffice.

<u>LECTURE 1</u>

I <u>Hydrocarbons</u>

   A. Alkanes: all single bonds; CH_4, C_2H_6, C_3H_8, - -

      1. Structural isomerism: same molecular formula but different arrangement of atoms. Two isomers of C_4H_{10}, three isomers of C_5H_{12}. Isomers of $C_3H_6Cl_2$:

$$
\begin{array}{cccc}
\underset{|}{Cl}\ \underset{|}{Cl} & \underset{|}{Cl} & \underset{|}{Cl} & \underset{|}{Cl}\qquad\underset{|}{Cl}\\
C - C - C & C - C - C & C - C - C - Cl & C - C - C\\
& \underset{|}{\ }\\
& Cl
\end{array}
$$

      2. Nomenclature

         a. Straight-chain: methane, ethane, propane,

         b. Branched-chain

216

suffix: find longest continuous carbon chain
prefix: methyl, ethyl, etc. Use number to indicate where branch-
ing occurs.

$$H_3C - \overset{\overset{\displaystyle CH_3}{|}}{\underset{\underset{\displaystyle CH_3}{|}}{C}} - \overset{\overset{\displaystyle H}{|}}{\underset{\underset{\displaystyle H}{|}}{C}} - CH_3 \qquad\qquad \text{2,2-dimethylbutane}$$

B. Alkenes: one double bond; C_2H_4, C_3H_6

Geometric isomerism: cis and trans

$$\begin{array}{ccc}
\underset{Cl}{\overset{Cl}{\diagdown}}C = C\underset{H}{\overset{H}{\diagup}} &
\underset{H}{\overset{Cl}{\diagdown}}C = C\underset{Cl}{\overset{H}{\diagup}} &
\underset{H}{\overset{Cl}{\diagdown}}C = C\underset{H}{\overset{Cl}{\diagup}} \\
(1) & (2) & (3)
\end{array}$$

1 and 2 are structural isomers; 2 and 3 are geometric

LECTURE 2

I <u>Hydrocarbons</u>

Alkynes; derivatives of acetylene, C_2H_2

Aromatic: derivatives of benzene, C_6H_6

benzene: naphthalene:

interpretation of "circle" structures

Nomenclature of benzene derivatives (ortho, meta, para)

II <u>Functional Groups</u>

A. Alcohols: -OH group (CH_3OH, C_2H_5OH, two isomers of C_3H_7OH)
Ethanol is made by fermentation of sugars or by hydration of ethene:

$$C_2H_4(g) + H_2O(g) \longrightarrow C_2H_5OH(l)$$

Use in alcoholic beverages: % of alcohol varies from 4% in beer
to 40% in brandy. Proof = 2 x volume % alcohol

217

B. Acids: $- C - OH$ Formic acid: $HCOOH$ Acetic acid: CH_3COOH
$\quad\quad\quad\quad\; \parallel$
$\quad\quad\quad\quad\; O$

Treatment with base gives salt:

$$CH_3COOH(aq) + OH^-(aq) \longrightarrow CH_3COO^-(aq) + H_2O$$

Soaps are sodium salts of long-chain fatty acids:

$$CH_3(CH_2)_{16}COO^-, \; Na^+ \quad \text{sodium stearate, found in soap}$$

C. Esters: formed by reaction of alcohol with carboxylic acid

$$CH_3 - C - OH + HO - CH_3 \longrightarrow CH_3 - C - O - CH_3 + H_2O$$
$$\quad\quad \parallel \quad\quad\quad\quad\quad\quad\quad\quad\quad\quad \parallel$$
$$\quad\quad O \quad\quad\quad\quad\quad\quad\quad\quad\quad\quad O$$

$\quad$ acetic acid $\quad\quad$ methanol $\quad\quad\quad$ methyl acetate

Fats are esters of long-chain carboxylic acids with glycerol

$$R_1 - C - O - CH_2$$
$$\quad\; \parallel \quad\quad\quad |$$
$$\quad\; O \quad\quad\quad |$$
$$R_2 - C - O - CH$$
$$\quad\; \parallel \quad\quad\quad |$$
$$\quad\; O \quad\quad\quad |$$
$$R_3 - C - O - CH_2$$
$$\quad\; \parallel$$
$$\quad\; O$$

If R group contains a double bond, fat is unsaturated

<u>LECTURE $2\frac{1}{2}$</u>

I <u>Polymers</u>

A. Addition polymers. Alkene or derivative of alkene adds to itself
to form a long-chain polymer. Polyethylene (from C_2H_4):

```
    H   H   H   H   H   H   H   H
    |   |   |   |   |   |   |   |
  - C - C - C - C - C - C - C - C-        may contain 2000 or more C2H4
    |   |   |   |   |   |   |   |         units
    H   H   H   H   H   H   H   H
```

may contain 2000 or more C_2H_4 units

Polyvinyl chloride, made from C_2H_3Cl:

218

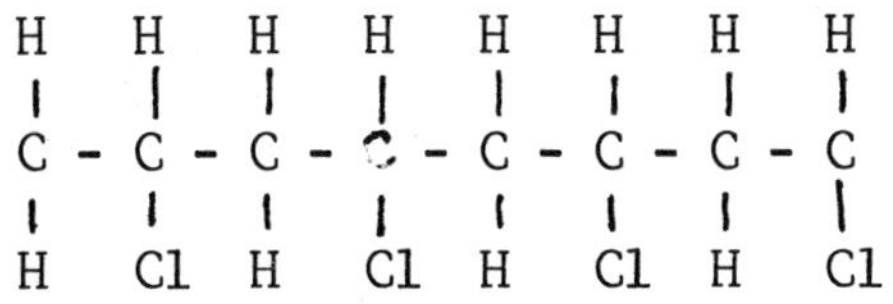

H - C - C - C - C - C - C - C - C   "head-to-tail" polymer

(polyvinyl chloride chain with H and Cl substituents)

B. Condensation polymers

   1. Polyesters: made from dicarboxylic acid + dialcohol

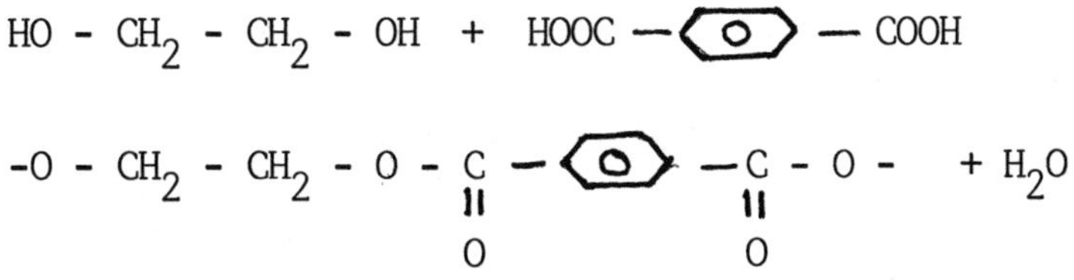

$$HO - CH_2 - CH_2 - OH \; + \; HOOC - \langle O \rangle - COOH$$

$$-O - CH_2 - CH_2 - O - \underset{O}{\overset{\|}{C}} - \langle O \rangle - \underset{O}{\overset{\|}{C}} - O - \; + \; H_2O$$

dacron

   2. Polyamides: made from dicarboxylic acid and diamine

$$H_2N - (CH_2)_6 - NH_2 \; + \; HOOC - (CH_2)_4 - COOH$$

$$- \underset{H}{\overset{|}{N}} - (CH_2)_6 - \underset{H}{\overset{|}{N}} - \underset{O}{\overset{\|}{C}} - (CH_2)_4 - \underset{O}{\overset{\|}{C}} - \; + \; H_2O$$

nylon

DEMONSTRATIONS

1. Preparation of acetylene: GILB A 35; J. Chem. Educ. <u>71</u> 253 (1994)
2. Preparation of ethanol by fermentation: GILB N 18
3. Esterification: GILB N 19
4. Nylon rope trick: GILB P 2; SHAK <u>1</u> 213
5. Different forms of polyethylene: GILB P 17
6. Bakelite formation: GILB P 11
7. Urea-formaldehyde plastic: SHAK <u>1</u> 227

<u>VIDEO</u>

1. Preparation of nylon: SHAK 15

Many additional demonstrations in polymer chemistry are available
from the POLYED National Information Center, University of Wisconsin-
Stevens Point, Department of Chemistry, Stevens Point WI 54481-3897

PROBLEMS

1. a. 2-methylbutane   b. 3-methylpentane

    c. 2,3-dimethylpentane   d. 2,3,5-trimethylhexane

```
                                  C
                                  |
3.  a.  C - C - C - C - C    b.  C - C - C - C
                |                    |
                C                    C
                |
                C

    c.  C - C - C - C - C - C - C   d.  C - C - C - C - C
            |   |                           |   |
            C   C                           C   C
                |
                C

5.  a.  C - C - C - C - C - C - C - C        4-isopropyloctane
                    |
                C - C - C

    b.  C - C - C        2-methylbutane
            |
            C
            |
            C

    c.  C - C - C        2-methylbutane
        |   |
        C   C
```

7. a. 2-methyl-1-propene   b. 2,3-dimethyl-2-butene

    c. 2-pentene   d. 2-methyl-1-butene

9. a. m-dibromobenzene   b. o-dinitrobenzene   c. p-diiodobenzene

11. a. methanol, methyl alcohol   b. ethanoic acid, acetic acid
     c. 1-propanol, propyl alcohol

13. a.   H – C – O – CH$_3$ b.   CH$_3$ – CH$_2$ – C – O – CH$_2$ – CH$_3$
             ‖   ‖
             O   O

                                                                    CH$_3$
                                                                    |
    c.   CH$_3$ – C – O – CH$_2$ – CH$_2$ – CH$_3$ d.   CH$_3$ – C – O – C – CH$_3$
               ‖   ‖ |
               O   O   H

                                    C   C
                                    | |
15. C – C – C – C – C – C ;   C – C – C – C – C ;   C – C – C – C – C ;

          C   C   C
          | | |
     C – C – C – C ;   C – C – C – C
          |
          C

                                                              Cl
                                                              |
17. C – C – C – C – Cl ;   C – C – C – C ;   C – C – C
                                       | |
                                       Cl   C

          C
          |
     C – C – C – Cl

19. (benzene ring with Cl and Br in adjacent positions) ; (benzene ring with Cl and Br in meta positions) ; (benzene ring with Cl and Br in para positions)

                                                              C
                                                              |
21. C = C – C – C – C ;   C – C = C – C – C ;   C = C – C – C ;

          C   C
          | |
     C – C = C – C ;   C – C – C = C

23. C – C – C – C – C – OH;   C – C – C – C – C;   C – C – C – C – C
                                          | |
                                          OH   OH

23. (cont.)

$$\begin{array}{ccccc} & & \overset{\textstyle C}{\overset{|}{}} & & \\ HO - & C - & C - & C - C; \end{array}$$

$$\begin{array}{ccccc} & & \overset{\textstyle C}{\overset{|}{}} & & \\ C - & C - & C - & C; \\ & & \underset{|}{} & & \\ & & OH & & \end{array}$$

$$\begin{array}{ccccc} & & \overset{\textstyle C}{\overset{|}{}} & & \\ C - & C - & C - & C \\ & & \underset{|}{} & & \\ & & OH & & \end{array}$$

$$\begin{array}{ccccc} & & \overset{\textstyle C}{\overset{|}{}} & & \\ C - & C - & C - & C; \\ & & \underset{|}{} & & \\ & & OH & & \end{array}$$

$$\begin{array}{ccc} & \overset{\textstyle C}{\overset{|}{}} & \\ C - & C - & C - OH \\ & \underset{|}{} & \\ & C & \end{array}$$

25. CH_3, H, $C = C$, H, C_2H_5 and CH_3, C_2H_5, $C = C$, H, H

27. $HOOC$, $COOH$, $C = C$, H, H $HOOC$, H, $C = C$, H, $COOH$

malic   fumaric

29. a. alcohol   b. ester, acid   c. alcohol, acid

31. a. $C - C - C - C$ b. $C - C - C - C - COOH$
              $\underset{|}{}$
              OH

c. $CH_3 - \overset{\overset{\textstyle H}{|}}{\underset{\underset{\textstyle C_2H_5}{|}}{C}} - O - \overset{\overset{}{}}{\underset{\underset{\textstyle O}{\|}}{C}} - C_4H_9$

33. a. $-\overset{\overset{\textstyle Cl}{|}}{\underset{\underset{\textstyle Cl}{|}}{C}} - \overset{\overset{\textstyle Cl}{|}}{\underset{\underset{\textstyle Cl}{|}}{C}} - \overset{\overset{\textstyle Cl}{|}}{\underset{\underset{\textstyle Cl}{|}}{C}} - \overset{\overset{\textstyle Cl}{|}}{\underset{\underset{\textstyle Cl}{|}}{C}} -$

b. $3.2 \times 10^3 \times 165.82 \text{ g/mol} = 5.3 \times 10^5 \text{ g/mol}$

c. CCl_2;   C: $\dfrac{12.01}{82.91} \times 100\% = 14.49\%$ $Cl = 85.51\%$

35. a.
$$
\begin{array}{cccc}
H & H & H & H \\
| & | & | & | \\
- C - C - C - C - \\
| & | & | & | \\
H & CN & H & CN
\end{array}
$$
b.
$$
\begin{array}{cccc}
H & H & H & H \\
| & | & | & | \\
- C - C - C - C - \\
| & | & | & | \\
H & CN & CN & H
\end{array}
$$

37. $H_2C = CHCl$ and $H_2C = C\begin{smallmatrix}H\\ \diagdown\end{smallmatrix}$ (phenyl)

39.
$$
\begin{array}{cc}
O - C - C - O - CH_2 - CH_2 - O - \\
\;\;\;\; \| \;\;\; \| \\
\;\;\;\; O \;\;\; O
\end{array}
$$

41.
$$
H_2N - CH_2 - C - OH \\
\;\;\;\;\;\;\;\;\;\;\;\;\;\;\; \| \\
\;\;\;\;\;\;\;\;\;\;\;\;\;\;\; O
$$

43. a. 109.5°   b. 109.5°, 120°   c. 109.5°, 180°

45. Three pairs of electrons spread around ring

47. a. reaction of CO with H_2 b. hydration of ethene

    c. oxidation of ethanol

    d. reaction of ethanol with acetic acid

49. a. no multiple bonds

    b. sodium salt of long-chain carboxylic acid

    c. twice the volume percent of alcohol

    d. ethanol with additives that make it unpalatable

51. a.
$$
\begin{array}{cc}
Br & H \\
| & | \\
H - C = C - H
\end{array}
$$
b.
$$
\begin{array}{c}
H \\
| \\
H - N - CH_2 - COOH
\end{array}
$$

    c. $HOOC - (CH_2)_4 - COOH$ and $H_2N - (CH_2)_6 - NH_2$

52. $C_{27}H_{46}O$

53. C - C - C - C - C - C - OH ;   C - C - C - C - C - C
                                                      |
                                                      OH

        OH   C
        | |
C - C - C - C - C - C ;   HO - C - C - C - C - C

        OH   OH   OH
        | | |
C - C - C - C - C ;   C - C - C - C - C ;   C - C - C - C - C
    | | |
    C   C   C

C - C - C - C - C - OH ;   HO - C - C - C - C - C
    | |
    C   C

        OH   OH   C - OH
        | | |
C - C - C - C - C ;   C - C - C - C - C ;   C - C - C - C - C
        | |
        C   C

        C   C   C
        | | |
HO - C - C - C - C ;   C - C - C - C ;   C - C - C - C - OH
        | | | |
        C   C   OH   C

                              OH
                              |
HO - C - C - C - C ;   C - C - C - C ;   17 in all
        | | | |
        C   C   C   C

54. $C_3H_8(g) + 5\ O_2(g) \longrightarrow 3CO_2(g) + 4H_2\ (1)$

     $\Delta H° = 3(-393.5\ kJ) + 4(-285.8\ kJ) + 103.8\ kJ = -2219.9\ kJ$

     $q_{water} = 1.00\ qt \times \dfrac{1\ L}{1.057\ qt} \times \dfrac{1000\ g}{1\ L} \times 4.18\ \dfrac{J}{g\cdot°C} \times 75.0°C$

     $= 2.97 \times 10^5\ J = 297\ kJ$

     mass propane = 297 kJ x (44.09 g/2219.9 kJ)   6 g

     (neglecting the heat capacity of the saucepan and assuming liquid
     water is produced)

224

Saunders College Publishing

55. a.

$$O - CH_2 - \underset{\underset{\displaystyle OH}{|}}{\overset{\overset{\displaystyle H}{|}}{C}} - CH_2 - O - \underset{\underset{\displaystyle O}{\|}}{C} \hspace{2em} \underset{\underset{\displaystyle O}{\|}}{C} - O - CH_2 - \underset{\underset{\displaystyle OH}{|}}{\overset{\overset{\displaystyle H}{|}}{C}} - CH_2 - O -$$

b.

$$- O - CH_2 - \underset{O}{\overset{\overset{\displaystyle H}{|}}{C}} - CH_2 - O - \underset{\underset{\displaystyle O}{\|}}{C} \hspace{2em} \underset{\underset{\displaystyle O}{\|}}{C} - O - CH_2 - \underset{O}{\overset{\overset{\displaystyle H}{|}}{C}} - CH_2 - O -$$

$$C = O \hspace{8em} C = O$$

$$C = O \hspace{8em} C = O$$

$$- O - CH_2 - \underset{\underset{\displaystyle H}{|}}{C} - CH_2 - O - C \hspace{2em} C - CH_2 - O - \underset{\underset{\displaystyle H}{|}}{C} - CH_2 - O -$$